MANUAL FOR CLIMATE CHANGE ADAPTATION MEASURES FOR TRANSPORT INFRASTRUCTURE IN CENTRAL ASIA WITH A FOCUS ON UZBEKISTAN

DECEMBER 2023

ASIAN DEVELOPMENT BANK

ADB

Contents

Tables, Figures, and Box

Tables

Figures

Box

Scope and Structure of the Manual

This manual aims to provide government officials, consultants, and contractors with useful information and resources on the impact of climate change in Central Asia and to show how adaptation measures can increase the resilience of the built transport infrastructure.

Considering in particular Kazakhstan, Kyrgyz Republic, Tajikistan, Turkmenistan, and Uzbekistan, the manual discusses the practical effects of climate change on transport infrastructure. It shows that understanding the potential threat posed by climate change is essential to ensuring that infrastructure is designed and built to be more resilient.

In this manual, transport infrastructure refers to essential components such as asphalt and concrete pavement, embankments, bridges, slopes, trenches, and drainage systems. Railways share similar structural solutions with motorways (drainage, trenches, embankment, bridges, tunnels), as well as unique elements such as tracks, cables, and signaling. Airport pavement, which suffers the same distresses as flexible (asphalt) and rigid (concrete) road pavement, is also among the structures that are vulnerable to the impact of climate change.

The manual shows how Central Asia's core transport infrastructure is expected to be impacted by the intensifying droughts, extreme rain events, glacial melt, and desertification predicted by climate change models. It assesses the effects of gravity flows, avalanches, glacial and permafrost melt, and heat on pavement.

The manual also discusses general adaptation measures to improve the resiliency of infrastructure in the region and concludes with a section focusing on Uzbekistan. Principal climate change stressors are reviewed and discussed, together with adaptation measures, including the use of low-cost bioengineering solutions to reduce the vulnerability of infrastructure to extreme weather.

Abbreviations, Weights, and Measures

Abbreviations

ADB	Asian Development Bank
CCA	climate change adaptation
CFP	carbon footprint
CO_2	carbon dioxide
EWS	early warning system
GEV	generalized extreme value
GHG	greenhouse gas
IPCC	Intergovernmental Panel on Climate Change
SLS	serviceability limit state
ULS	ultimate limit state

Weights and Measures

GPa	gigapascal (1 GPa = 1,000 MPa)
kg	kilogram
km	kilometer
km^2	square kilometer
Kt	kiloton
m^3	cubic meter
mm	millimeter
MPa	megapascal (1 MPa = 10 kg/square centimeter)
ppm	part per million

Acknowledgments

This publication was prepared by Pawan Karki, principal transport specialist, Asian Development Bank (ADB); and Michel Di Tommaso, ADB consultant; supported by Hideaki Iwasaki, deputy director general of ADB's Pacific Department.

The authors would like to thank the following ADB staff—Arghya Sinha Roy, principal climate change specialist (climate change adaptation), and Michael Anyala, senior road asset management specialist—for their help in reviewing the report.

Basic Terminology and Concepts Related to Climate Change

1

A. Basic Terminology

We introduce here some important terms frequently used when addressing climate change.

Stressor: a climate-driven set of actions on a structure

Resilience (for roads): residual resistance after nth repetition of a stressor

Carbon footprint (CFP): embodied carbon in a product and/or process

Mitigation: actions to reduce CFP of a product and/or process

Adaptation: actions required to enhance resilience

Paris Accords (2015): agreements sanctioned by more than 190 nations worldwide to maintain an average temperature of less than 2.0°C of pre-industrial levels by the end of the 21st century

CFP reduction goals: to reduce CFP by 50% by 2030 and to be carbon-neutral by 2050

Representative Concentration Pathway (RCP): scenarios for climate change, ranging from 2.6 watts per square meter (RCP2.6) to 8.5 watts per square meter (RCP 8.5) by the end of the 21st century

Emitter: process (natural or artificial) that produces carbon dioxide

Sink: natural or artificial process that absorbs carbon dioxide

Carbon neutrality: condition wherein sinks balance emitters

B. Introduction

It is a well-established fact that atmospheric carbon dioxide (CO_2) remained stable at about 280 parts per million (ppm) throughout the Holocene period.[1] From around 1840 to the 1950s, concentration grew from 280 ppm to about 310 ppm. Since the 1950s, it has grown at an accelerated pace to reach present-day values (about 410 ppm and still growing). CO_2 and other greenhouse gases (GHGs), such as methane and water vapor, are responsible for creating a beneficial greenhouse effect wherein a stable concentration of these gases is needed to maintain human life on Earth. However, when the concentration of these gases increases (mainly because of human activities), more heat is trapped in the system than what can be radiated back to space, causing a slow average temperature rise, which is called global warming.

To keep the average global temperature at less than 2.0°C above pre-industrial levels by year 2100 (with a safe maximum target of 1.5°C), current climate models indicate that the CO_2 from anthropogenic emissions should be reduced by half by 2030, and that the planet should become carbon-neutral by 2050. This goal was sanctioned by 193 countries adhering to the Paris Agreement 2015.

Failing to meet these boundaries would mean reaching a potential tipping point, wherein the irreversible, unpredictable, and nonlinear responses of the climate system could materialize, and whereby current models would be unable to make accurate predictions.[2]

Falling short of the goals set in the Paris Agreement 2015 of limiting global temperature rises to 1.5°C above pre-industrial levels risks throwing the climate system into chaos.[3] In other words, what today are seen as *worst-case scenarios,* based on current climate models, may become *best-case scenarios* in the near future. Such events may include the irreversible reduction of human habitat due to factors such as extreme heat, sea level rise, increasingly intense weather events in coastal areas (tidal surges, hurricanes, etc.), and extreme geohazards, among others. In terms of the impact of climate change on transport infrastructure, the intensification of weather events, both in frequency and intensity, will be a primary concern because of the potential damage and safety risks for users. Extreme weather events, which were considered unlikely a few decades ago and with very long return periods, are becoming more likely and frequent on a yearly basis. With weather extremes occurring more often as a consequence of climate change, infrastructure will become more vulnerable to damage, increasing the probability of premature failure of bridges, embankments, and slopes, among others. Some of the most vulnerable parts of the transport infrastructure to climate-driven events are the pavement structures (concrete or asphalt road), embankments and trenches (slopes), bridges, overpasses, underpasses, tracks, tunnel portals, and drainage systems.

---

[1]   This is in reference to the period for the last 12,000 years and continuing to the present.
[2]   A. Fares, ed. 2021. *Climate Change and Extreme Events.* Elsevier.
[3]   In mathematical terms, little variations in the boundary conditions (input) lead to very large variations in the weather events (output).

The effects of extreme climate events on transport infrastructure. Core transport infrastructure in the region is vulnerable to the impacts of climate change (photos by Pawan Karki and Michel Di Tommaso).

C. Climate Change in Central Asia

The Central Asian region, which includes Kazakhstan, Kyrgyz Republic, Tajikistan, Turkmenistan, and Uzbekistan, is already feeling the growing impacts of climate change. Going forward, the region can expect to face climate stressors including temperature rises, extreme weather events, and increasing glacial melt, alongside the continued expansion of deserts and arid areas.[4]

With a population of about 65 million, most of the Central Asian region falls within arid and semiarid zones and is covered by grasslands, deserts, and woodlands, with a dramatic physical landscape ranging from grassy steppes and high mountains to deserts and large rivers, lakes, and seas.

Reported average annual temperature increases since the 1950s vary widely—for example, from 0.3°C to 1.2°C in Tajikistan, and from 1.1°C to 2.4°C in Turkmenistan and Uzbekistan.[5] The per-decade increase in average annual temperatures has increased significantly. For example, Tajikistan has seen an overall 0.1°C per-decade increase from 1901 to 2013 to 0.4°C per decade over the past 30 years.

---

4   Intergovernmental Panel on Climate Change (IPCC). 2022. *Climate Change 2022: Impacts, Adaptation and Vulnerability*. Contribution of Working Group II to the Sixth Assessment Report of the Intergovernmental Panel on Climate Change (H. O. Pörtner, et al., eds.). Cambridge, UK and New York: Cambridge University Press.
5   Climate Links.

In the Kyrgyz Republic, it has shifted from an overall 0.17°C per-decade increase from 1901 to 2013, to around 0.5°C per decade over the past 30 years.

As a result of increasing temperatures, nearly one-third of glacial areas in the region have disappeared since 1930. Melting of permanent snowcaps and increases in temperature have also caused an increase of rockfalls—the Chamoli disaster took place in 2021 in the Nanda Devi National Park (India) where an abrupt collapse of a portion of a glacier caused, after detachment, the collapse of the flanks of the valley.

Across the Central Asian region, temperature increases are more pronounced at low altitudes and are most marked during the winter months, particularly in November and December. Over the past 30 years, total annual precipitation has risen by between 4% and 7% in Uzbekistan and the Kyrgyz Republic. Elsewhere, there have been slight decreases in Turkmenistan, and no clear trend determined in Kazakhstan and Tajikistan.

Projected changes in the region include the following (footnote 5):

- Likely increased average annual temperature of about 2.0°C by 2050, with reference to pre-industrial levels. For 2085, ranges could be from 2.0°C to 5.5°C in the Kyrgyz Republic and from 2.2°C to 5.7°C in Tajikistan.
- Greater storm intensity and frequency, alongside increased drought and prolonged dry spells.
- A higher frequency of heavy rain events. Precipitation, specifically for the November–April period, is expected to increase, with projected rises of up to 50% in the Kyrgyz Republic and up to 30% in Uzbekistan by 2085. Other months are likely to experience no change.

Useful materials, information, and projections can be found in the *Interactive Atlas* of the Intergovernmental Panel on Climate Change (IPCC).[6]

D.   Carbon Dioxide Emissions, Climate Change, and Global Warming

When dealing with climate change, emitters are defined as the combined sources of GHGs and sinks as the absorbers.

CO_2 emissions can be divided into two categories: those stemming from natural systems, and those caused by humans (anthropogenic emissions). Emissions from natural systems include forest fires, oceans, wetlands, mud volcanoes, volcanoes, and earthquakes. Anthropogenic emissions accounted for approximately half of the total global CO_2 emissions as of 2016.

Natural carbon sequestration through carbon sinks (i.e., trees, plants, and oceans) balances, roughly, the natural emissions only. This means that anything extra added to natural emissions makes the system no longer carbon-neutral, and it exerts extra pressure on what is an otherwise self-balancing earth system. Anthropogenic emissions are an "extra" input in the earth system.

---

[6]   IPCC. WGI Interactive Atlas.

With the recent increase in the rate and intensity of summer wildfires in the western United States, the Russian Federation's Siberia region, Southern Europe, and Australia, among others, the balance between natural emissions and natural sinks is disrupted even more, as more CO_2 is produced and sinks effectively become emitters.

Forest burning means that carbon sinks effectively become carbon emitters. With the reduction of forest lands and the increase of emissions from anthropogenic activities, oceans in turn absorb more CO_2. Some of the CO_2 absorbed by oceans is used by organisms to produce calcium carbonate (limestone) shells. The CO_2 concentration in the oceans needs to be such to favor the formation of shells, which are made of limestone ($CaCO_3$). If excessive CO_2 is absorbed by the world oceans, the end result is acidification caused by the increase in the concentration of carbonic acid (H_2CO_3).

E.   Anthropogenic Emissions

About 7,500 large CO_2 emitters have been identified,[7] but only a fraction of these contribute to roughly 99.6% of the total human-made emissions.

As shown in Figure 1, fossil fuel burning and mineral processing are by far the largest contributors to anthropogenic emissions at 78.8%, followed by cement production (7.0%), refineries (6.0%), and the iron and steel industry (4.8%).

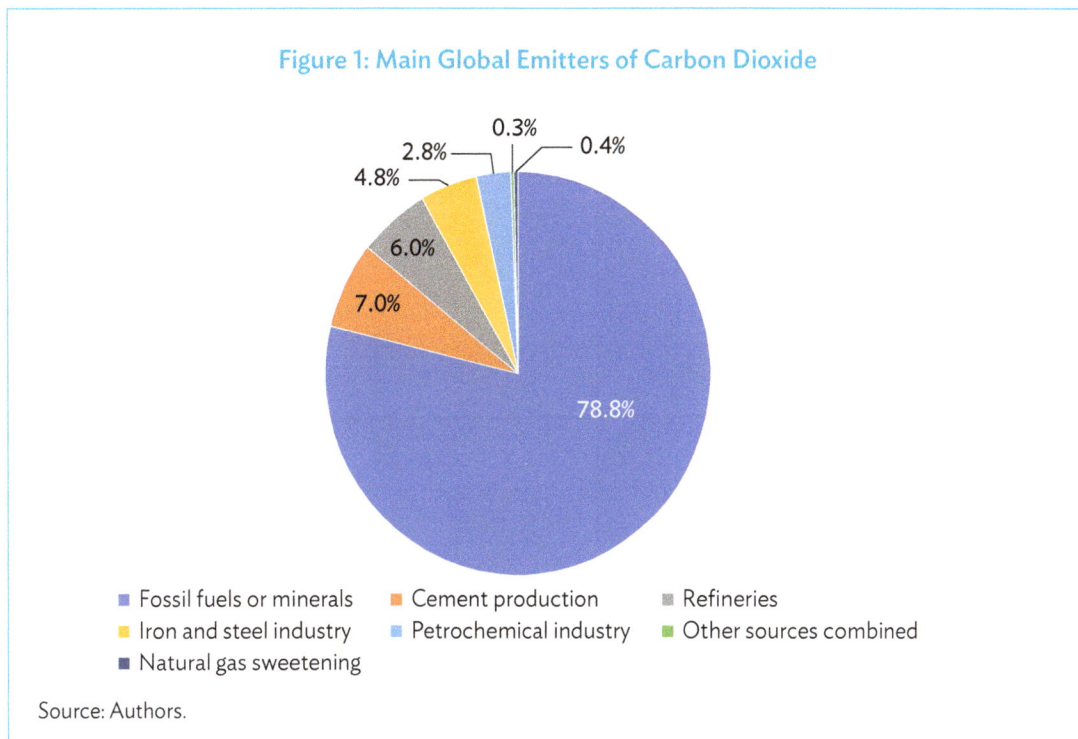

Figure 1: Main Global Emitters of Carbon Dioxide

- Fossil fuels or minerals
- Cement production
- Refineries
- Iron and steel industry
- Petrochemical industry
- Other sources combined
- Natural gas sweetening

Source: Authors.

---

[7]   Large emitter means emissions above 0.1 million tons CO_2 per year.

F.   Emitters and Sinks

GHGs are those gases in the atmosphere, such as water vapor, methane, and CO_2, that help trap part of the radiated heat produced by the Earth daily. In doing so, they keep the surface temperature at levels that can sustain human life at almost all latitudes and elevations. Sinks and emitters may be natural or anthropogenic.

Natural carbon sinks are trees, for instance, consuming CO_2 to produce oxygen as a by-product of their living cycle, while an artificial carbon sink is, for example, a thermally cracked char. The global yearly quantity of dead wood in our planet is roughly at 67 gigatons of biomass, which is about 11% of all biomass.[8] A vast amount of this dead biomass is currently burned in open air, particularly in developing countries, leading to a large-scale (regional) phenomenon of polluting haze, rich in CO_2. If this biomass is subjected to combustion without oxygen (pyrolysis), the end product fixes the carbon in the organic matter, preventing it from reacting with atmospheric oxygen. No extra CO_2 is produced as a consequence of this process.[9] This is an example of using a natural organic material with a high potential release of CO_2, if burned conventionally, to convert it into another manufactured organic material, which can be used in many applications, from food to agriculture to filters, without producing extra CO_2.

Because we cannot control natural emitters of GHGs such as volcanoes, we cannot mitigate their impact on climate change. Exceptional volcanic events may add millions of tons of GHGs in the atmosphere in a few hours, causing climate to vary for decades for some regions, or even for the entire globe. We are, in fact, defenseless against the impact of these natural events on climate change. We can, however, act on human-made emissions by reducing them and preventing the irreversible changes in our climate system. But, even if we cut by half now all human-made CO_2 emissions, we would still face the effects of climate change for some more decades because, before the climate system would revert toward pre-industrial global warming levels, the planet's average temperature will rise for some time because of its thermal inertia. Hence, together with doing our best to reduce GHGs through mitigation, we also need to realistically be prepared by adapting to a major impact of climate change on the transport infrastructure.

---

[8]   M. Di Tommaso. 2022. *Making the Unsustainable Sustainable: What Lies Ahead in the Concrete Industry in View of the Paris 2015 Agreements on Climate* (published on LinkedIn).

[9]   M. Di Tommaso. 2016. *NOx Adsorption, Fire Resistance and CO_2 Sequestration of High Performance, High Durability Concrete Containing Activated Carbon.* Paper presented at the Second International Conference on Concrete Sustainability - ICCS16. Madrid. 13–15 June.

Mitigation

2

A. Introduction

A two-pronged approach focuses on (i) effectively reducing the carbon footprint (CFP) of human-related activities on the environment, and (ii) bracing for the future impacts of a changing climate.

Mitigation aims to reduce as much as possible the impact of human activities on the climate system, and to modulate the levels of GHGs so that the ecosystems and the environment become, once again, sustainable. It focuses on reducing the production of heat-trapping GHGs by (i) reducing the sources of these gases (fossil fuels or electricity and heat generation); (ii) using sustainable materials with a low CFP from "cradle to gate";[10] and (iii) enhancing the "sinks" that capture and accumulate GHGs (such as afforestation along roads, production of reusable thermally cracked biochar, and the recycling of construction materials).

Adaptation requires—in the context of transport infrastructure—finding ways to cope with the increased risk of premature failure caused by extreme events and changed environmental conditions. Measures can be taken to extend the service life of structures by implementing redundancy in designs and environmentally sustainable solutions.

With the assistance of the Asian Development Bank (ADB), adaptation in Central Asia has been central to developing an internationally benchmarked guidance and good practice for addressing the impact of climate change and for building resiliency in the transport sector. This upstream support aims to enhance strategic transport planning, guidelines, and standards, and to align them with the "build back better" objectives. This is defined as recovery that leads to improvements above and beyond the predisaster status quo.[11] ADB's Disaster and Emergency Assistance Policy uses the United Nations' definition of build back better in the document, *Sustainable Development: Disaster Risk Reduction,* which states that build back better entails recovery, rehabilitation, and reconstruction phases after a disaster to increase the resilience of nations and communities.[12] This is achieved through integrating disaster risk reduction measures into the restoration of physical infrastructure and societal systems, and into the revitalization of livelihoods, economies, and the environment. The policy, which is to be implemented during recovery and reconstruction phases, recognizes a window of opportunity to rebuild assets and improve livelihoods in ways that reduce the potential impact of future hazards.

---

[10]   "Cradle" refers to the manufacturing point, while "gate" refers to the delivery point.

[11]   B. Noy et al. 2019. Build Back Better: What Is It, and What Should It Be? *ADB Economics Working Paper Series.* No. 600. Manila: ADB.

[12]   ADB. 2019. *Review of the 2004 Disaster and Emergency Assistance Policy.* Manila; and United Nations General Assembly. 2016. *Report of the Open-Ended Intergovernmental Expert Working Group on Indicators and Terminology Relating to Disaster Risk Reduction.* Seventy-First Session, Agenda Item 19 (c).

How to Effectively Pursue Mitigation and Adaptation

Mitigation is mainly dealt with at the government level, where policymakers should consider introducing measures including

(i)   mandating afforestation to create natural sinks;[13]
(ii)   using low CFP materials and solutions to build new, and to refurbish existing, infrastructure;
(iii)   promoting, endorsing, and valuing sustainable designs and for construction; and
(iv)   enforcing regulations on maximum GHG emissions made by industrial processes, such as cement manufacturing and transport (e.g., fuel efficiency).

The above and similar measures are all required to limit the increase in temperature within sustainable values. However, even if emissions are reduced now, the climate will keep changing over the next few decades. Before things get better, they are likely to worsen.

Hence, adaptation is essential to ensure that new and existing infrastructure are built and reinforced to make them more resilient to the impacts of climate change in the near future.

Governments and engineering associations in charge of issuing specifications for design, construction, and rehabilitation of transport infrastructure must understand and quantify how these changes will impact existing and newly built infrastructure. In the United States (US), Europe, and Australia, for example, international standards and organizations, along with engineering institutes, play a contributing role. It is expected that all developing countries will also follow this example and introduce more environment-friendly policies related to design and construction.

B.   Carbon Footprint Calculations

In the last 20 years, institutions and governments worldwide have prepared databases of embodied carbon in construction materials. One of these tools is the *Inventory of Carbon and Energy,* published by the University of Bath in England.[14] The inventory contains a summary of about 1,800 records of embodied carbon in construction materials.[15] The database has been developed following the guidelines of the standard *ISO 14044:2006 Environmental Management—Life Cycle Assessment: Principles and Framework.*[16] The aim of such inventories is to allow engineers, contractors, and suppliers to determine how much CO_2 is embodied in some of the most common construction materials and processes, and to find ways to reduce it to meet the goals of the 2015 Paris Accords. Estimating the CFP of construction materials and processes in their first-time use is one of the mitigation measures required to reduce carbon emissions. Calculating the embodied carbon in construction processes is needed to implement

---

13   Afforestation is the action of planting new trees in an area where trees have been removed either by natural events (fires, for instance) or by human activities.
14   Greenhouse Gas Protocol.
15   Circular Ecology.
16   International Organization for Standardization (ISO). 2006. *ISO 14044:2006 Environmental Management—Life Cycle Assessment: Principles and Framework.* Geneva. The first edition of ISO 14044, together with ISO 14044:2006, cancels and replaces ISO 14040:1997, ISO 14041:1998, ISO 14042:2000, and ISO 14043:2000, which have been technically revised.

the mandatory actions toward the use, from now on, of sustainable designs and construction materials. The next two sections present two examples of basic mitigation measures, which should be a part of the evaluation of sustainability and/or resilience of transport infrastructure projects. The first example shows how to calculate the CFP of a concrete pavement, while the second example shows the importance of afforestation to offset a part of the CFP associated with the life cycle of a road.

Carbon Footprint of a Concrete Pavement

Industrial processes associated with production, delivery, construction, and demolition, among others, produce GHGs (CO_2) via the burning of fossil fuels. CFP is the summation of CO_2 emissions created by the manufacture, transport, construction, and demolition, among others, of all of the materials used to build transport infrastructure. Embodied carbon is usually expressed in kilogram (kg)-CO_2 per kg of construction material. For instance, producing Portland cement (with a high clinker factor of more than 95%) by burning limestone and clay above 1,600°C in the cement kilns produces approximately 0.90 kg-CO_2 per kg of cement.[17] Therefore, the CFP of a medium-strength concrete mix designed with 390 kg per cubic meter (m^3) of cement for a road pavement would have about 350 kg-CO_2 per m^3, without considering batching, transport, and vibration, among others. The utilization of by-products (limestone, slag, volcanic ash, etc.) replacing clinker reduces the CFP of cement by 8–9 kg-CO_2 per 1% reduction in the clinker factor. For example, incorporating limestone cement (which is Portland cement where some of the clinker is replaced by cold milled limestone) with a clinker factor of 75% reduces the CFP of medium-strength concrete to about 300 kg-CO_2 per m^3.

To form, for instance, a 50-kilometer (km)-long double-lane road (a 10-meter-wide concrete platform in total) with a 30-centimeter-thick concrete pavement, we need 50,000 m^3 of concrete. The CFP of the concrete batched is therefore 17.5 kiloton (Kt)-CO_2, neglecting the CFP of aggregates, which is very small compared to cement. Delivering and placing concrete to a distant location will add even more CO_2 because concrete is transported and vibrated using diesel-powered engines. Some figures of CFP for these processes are given below:

- Concrete plant: 0.18 kg-CO_2/m^3
- Truck mixer (agitator): 0.07 kg-CO_2/km
- Pump: 0.74 kg-CO_2/m^3
- Vibrator: 0.04 kg-CO_2/m^3

Going back to the example above, we would have additional 9.0 Kt-CO_2 from mixing at the plant to the 17.5 Kt-CO_2 embodied into the mixing process. For an average distance of, for example, 25 km between the plant and the site, and with concrete delivered in 10 m^3 trucks, it takes 5,000 trips (or 125,000 km) for the truck mixers to deliver the concrete to the site. Hence, the contribution of transport to the CFP of the project is another 8.7 Kt-CO_2. If concrete is vibrated, we consider finally another 2.0 Kt-CO_2.

The contribution to CFP of each of the emitters is given in Figure 2, which shows that the construction of concrete pavement has different components of CFP wherein the mix design, batching, and transport processes are dominant.

---

[17]   Limestone is made out of calcium carbonate ($CaCO_3$). Upon heating above certain temperature, it releases CO_2 to form calcium oxide (CaO), which is one of the key components of cement.

Figure 2: Carbon Footprint Contribution of Key Processes Related to Concrete Placement

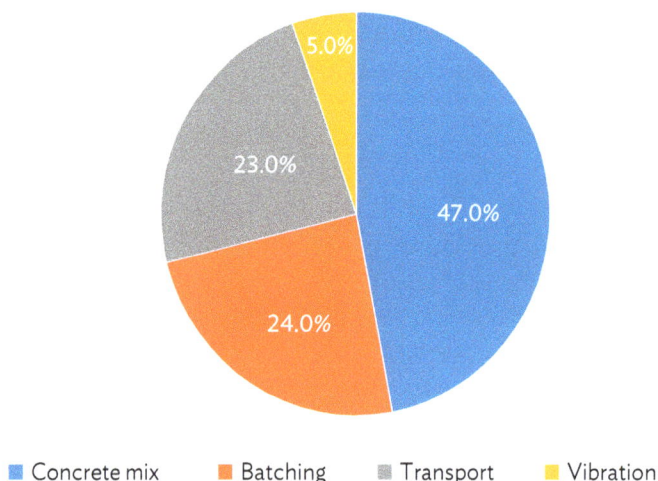

■ Concrete mix   ■ Batching   ■ Transport   ■ Vibration

Note: The concrete mix design and the batching of concrete have the largest carbon footprint among the concrete-related operations.
Source: Authors.

To reduce the CFP of a road project by 20%, for example, the designers must compare and adopt solutions that minimize the overall CFP of the construction process, such as

(i)   reducing the thickness of concrete pavement and/or using thicker stabilized subbases that can be produced with lower quantity of cement;

(ii)   using sustainable binders with lower CFP than Portland cement;

(iii)   using asphalt instead of concrete;

(iv)   reducing the travel distance between the plant and the site by installing an on-site batching plant; and

(v)   partially offsetting the embodied carbon in the construction materials used by adding trees and shrubs along the road, which are natural sinks.

All of the above are mitigation measures, and mitigation is only possible if we know the CFP of our activities and the ways to effectively reduce it to meet specific goals. This is what carbon databases are for and why their use should become more and more common among practitioners in the construction of transport infrastructure, among others. The CO_2 emissions related to construction activities shall be considered for the entire life span of the infrastructure—from production and maintenance to demolition and reuse (when possible)—and this can be done with appropriate tools, such as those described in the next sections of the manual.

Afforestation to Offset the Carbon Footprint of a Road

Afforestation involves planting trees and shrubs to populate areas where vegetation has been removed by human activity and/or it has been burned away by fires. Afforestation is both a mitigation and an adaptation strategy as trees and vegetation absorb CO_2 and stabilize slopes during rain events and provide other benefits. Because of their geographic features, countries like Uzbekistan and Turkmenistan are dominated by desert plants such as saxaul. One square kilometer (km^2) of land occupied by saxaul and shrubs can absorb about 3.3 Kt-CO_2 over a period of 20 years.

Planting shrubs and trees along a road does not merely provide a sink for the CO_2 produced by traffic on that specific area, it is also a mitigation measure to offset a part of the CFP embodied in the road itself. Afforestation should not be waived, but integrated into a broader strategy.

Vegetation also provides additional benefits such as stabilizing slopes against water saturation and collapse (see examples of bioengineering solutions in Chapter 4, section J on pages 55–56).

Using the road network to provide green pathways running parallel to the built transport infrastructure also creates benefits such as shaded walkways, cycleways, filtering of particulates, and habitat creation.

Over the last decade, many governments globally have made progress in adopting afforestation schemes as a mitigation measure. Morocco, for example, has planted about 3 million trees as of 2017 as part of the country's road afforestation projects. This is equivalent to sequestering about 300 Kt-CO_2 from the atmosphere in the next 10 years.

A useful mitigation tool is the CO_2 Calculator Spreadsheet (Figure 3) specifically developed by ADB to calculate the amount of CO_2 absorbed by trees and shrubs over a certain period of time (Appendix 1). The tool works in two ways—using the US Environmental Protection Agency method and the IPCC method.[18]

The example in Figure 3 shows the effect of planting 100 acres of shrubs over a 40-year design life for an infrastructure project in a semidesert area, calculated using the IPCC method. As 100 acres is 0.4 km^2, for a 25 km road, this means a stretch of 16-meter-wide shrubs.

Considering the life span of these species is a maximum of 7–10 years, periodic replenishing of dead plants is needed to maintain the efficiency of the system. One hundred hectares of shrubs may absorb up 6.5 Kt-CO_2 over 40 years.

Therefore, planting shrubs and desert trees along a 25 km concrete road would have an offset of -17% of the CFP embodied in the concrete road itself.

---

[18]   US Environmental Protection Agency.

Figure 3: Carbon Dioxide Calculator Spreadsheet

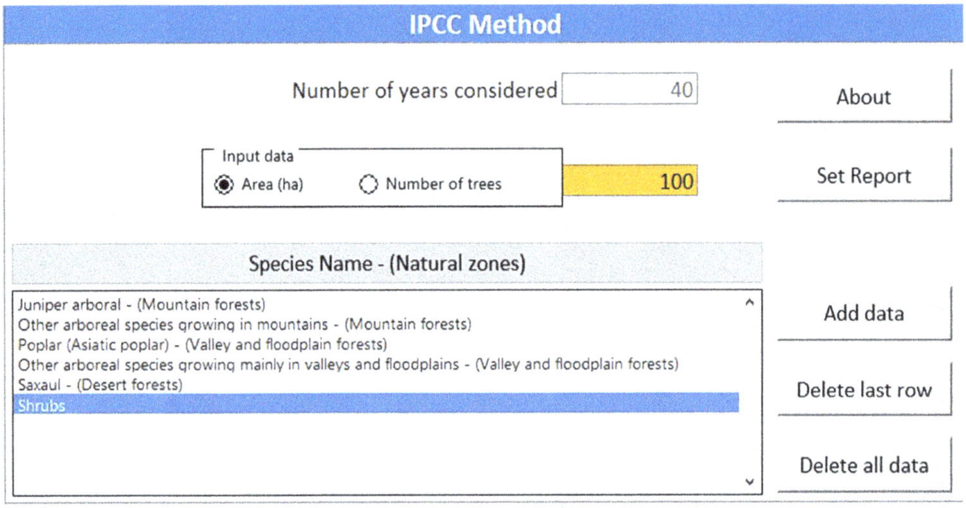

IPCC Method		
Number of years considered	40	About
Input data: ◉ Area (ha) ◯ Number of trees	100	Set Report

Species Name - (Natural zones)

Juniper arboral - (Mountain forests)
Other arboreal species growing in mountains - (Mountain forests)
Poplar (Asiatic poplar) - (Valley and floodplain forests)
Other arboreal species growing mainly in valleys and floodplains - (Valley and floodplain forests)
Saxaul - (Desert forests)
Shrubs

Add data

Delete last row

Delete all data

IPCC Method.   Number of years considered = 40   10/09/2022 - 08:34

Species Name	Number of trees estimated	Area (ha)	CO2 sequestred (t)
Shrubs	62,500	100.00	6,441.05
Total	**62,500**	**100.00**	**6,441.05**

CO_2 = carbon dioxide, ha = hectare, IPCC = Intergovernmental Panel on Climate Change, t = ton.

Note: The effect of planting 100 acres of shrubs over a 40-year design life for an infrastructure project in a semidesert area is calculated using the IPCC method. One hundred acres is 0.4 square kilometers or, for a 25-kilometer road, this means a stretch of 16-meter-wide shrubs. Considering the life span of these species is a maximum of 7–10 years, periodic replenishing of dead plants is needed to maintain the efficiency of the system. One hundred hectares of shrubs may absorb up 6.5 kilotons of CO_2 over 40 years. Hence, if we plant shrubs and desert trees along a 25-kilometer concrete road, we have an offset of -17% of the carbon footprint embodied in the concrete road itself.

Source: Asian Development Bank.

Climate-Related Stressors on Infrastructure: Causes and Effects

3

A. Introduction

Climate-related stressors contribute to weakening the built transport infrastructure such as roads and railways. This chapter considers floods and the effects of water saturation on the stability and bearing capacity of soils. It also looks at the damaging effects of flowing water and sediment load on bridges' decks, abutments, and piers. Gravity flow, the thermal effects on rigid and flexible pavements, and the impacts of snow accumulation and glacial melt are also assessed.

B. Floods: Risk Assessment

Flooding conditions are triggered by a combination of climate and weather-related factors (Figure 4).

Once flood conditions appear, the intensification of the phenomenon is governed by the morphology of the ground. Generally, there are two types of floods in Central Asia:

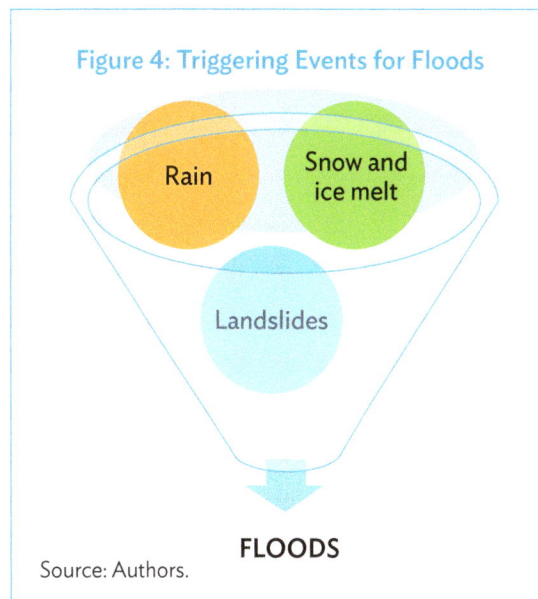

Figure 4: Triggering Events for Floods

Rain

Snow and ice melt

Landslides

FLOODS

Source: Authors.

(i) flash floods caused by unpredictable, quick, and abundant rainfall, combined with low ground permeability and specific topography (gorges, canyons, etc.); and

(ii) river floods, which occur when consistent rain or snow melt forces a river to overflow.

Based on current climate projections, these phenomena are expected to become more frequent and more intense in the region. In view of the above, and to elevate the resilience of the road infrastructure, the following should be considered:

• Evaluate—using adequate tools such as risk analysis and extreme value theory—the likelihood of certain extreme events to occur or reoccur during the service life of the infrastructure with higher probability than previously assumed.

- Plan and execute proper geological, geomorphological, and hydrological investigations to define areas that are at a higher risk of flooding than others and that may require specific measures.

C. Floods: Return Period and Generalized Extreme Value Distributions

Flood frequency analysis is a technique used by hydrologists to predict flow (or discharge rate) values corresponding to specific return periods along a river. Using annual maxima of flow data made available for a number of years, flood frequency analysis is used to calculate statistical information, such as mean and standard deviation, which is further used to create frequency distribution graphs. The frequency distribution (i.e., the model) is chosen from the available statistical distributions such as Gumbel, Weibull, and Fréchet (also known as the generalized extreme value or GEV). GEVs are useful tool for modeling extreme values, for instance, the maximum height of water in a river, the maximum height of waves on a coastal area, or the maximum wind speed recorded daily in a region over the past 50 years. This hydrological approach could address the following questions:

- Given the distribution of maxima of the height of water or discharge rate of a river in a given period, how will the maxima distribute over the next 10, 100, and 200 years?
- Considering the additive effects of climate change, how likely and frequent will those "rare events" recorded in the past reoccur?

Figure 5 shows an example of statistics of extremes or statistics of maxima to determine the probability of rare events, over scales of tens of years, to occur yearly.

Also, in hydrology, after selecting the GEV distribution that best fits the annual maxima data, flood frequency curves are plotted. Flood frequency plays a key role in providing estimates of recurrence of floods, which are used in designing structures such as dams, bridges, culverts, levees, and motorways. To evaluate the optimum design specification for hydraulic structures, and to prevent overdesigning or underdesigning, it is imperative to apply statistical tools to create flood frequency estimates. These estimates are useful in providing a measurement parameter to analyze the damage corresponding to specific flow rates during floods. An accurate estimation of flood frequencies not only helps engineers in designing safer structures, but also safeguards against the potential economic losses of early deterioration. To understand how flood frequency analysis works, it is important to understand the concept of return period.

D. Return Period: Definitions

A return period is the inverse of the probability that an event will be exceeded in a given year. In general, a return period provides an estimate of the likelihood of an event, such as flood, occurring over 1 year of service of a structure. Return periods in hydrology simply provide an estimate of the probability of exceedance of a given flow in a given year. For example, if the 100-year return period flow value for a river is 1,000 m^3 per second, it means that there is a 1 in 100 (or 1%) chance for this value to be exceeded in any given year.

Figure 5: Number of Maximum Tides Recorded Yearly in Venice

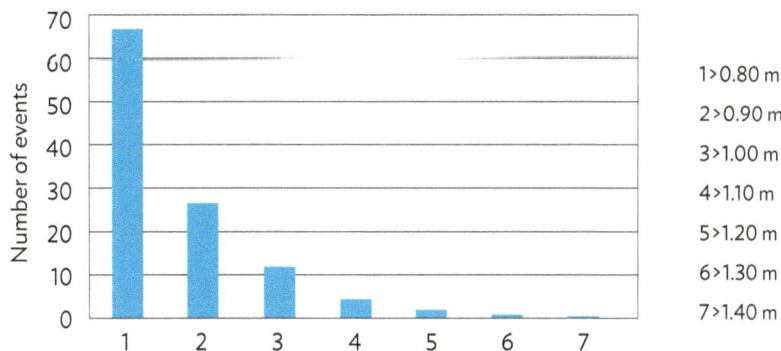

1	> 0.80 m
2	> 0.90 m
3	> 1.00 m
4	> 1.10 m
5	> 1.20 m
6	> 1.30 m
7	> 1.40 m

m = meter.

Notes:
1. The data are for the counts of the high tides in the city of Venice from 1966 to 2018.
2. Venice, as a UNESCO heritage site being vulnerable to rising sea levels, is a tangible example of the effects of climate change. We can observe that tides between +80 centimeters (cm) and +90 cm over the reference datum have been quite common, with multiple events per year over a period of 52 years; whereas, tides of more than 150 cm have been much less common and, in fact, very rare. If exceeding—for instance, 150 cm would mean catastrophic damage to the city (flooding of museums and historical buildings, damage to property, etc.)—we ask ourselves: how can we predict the likelihood of exceeding certain levels (e.g., 150 cm) in the future based on past records?

Source: Authors' elaboration based on Statista.

A flood frequency curve is used to relate flood discharge values to return periods to provide an estimate of the intensity of a flood event. The discharges are plotted against return periods using either a linear or a logarithmic scale. To provide an estimate of the return period for a given discharge rate of a river, the observed data are fitted with a theoretical distribution using a cumulative density function. Figure 6 shows that, if the slope of the best fit line produces a certain model of distribution of values for the past 50 years, the same model can be extrapolated to the future to estimate longer return periods. Therefore, the probability and magnitude of possible, yet unlikely, disruptive events to happen in any given year can be predicted.

It is important to predict the likelihood of an extreme event during the service life of a structure, and to estimate the consequences of the event when it happens. For example, the consequence of a river's height exceeding a certain maximum value could be the collapse of a bridge. Estimating the risk of a damaging event involves knowing how the height of a river has varied over time in the past, and how frequent the maxima are.

Return period and design or service life are complementary but different concepts. Service life is the time required—with properly conducted ordinary maintenance—to reach a certain serviceability limit state (SLS). For example, durability design of reinforced concrete requires the probability of initiation of corrosion induced by carbonation and/or by chloride to be equal to or less than 10% at any given year. When the condition is reached, service life is concluded, and maintenance is required to avoid quick deterioration and collapse. Finally, to further reduce the risk of the bridge's deck being washed away by an intense flood, it is important to add structural details, such as shaping the deck, to provide less hydrodynamic resistance to flow.

Figure 6: Flood Maxima Plot

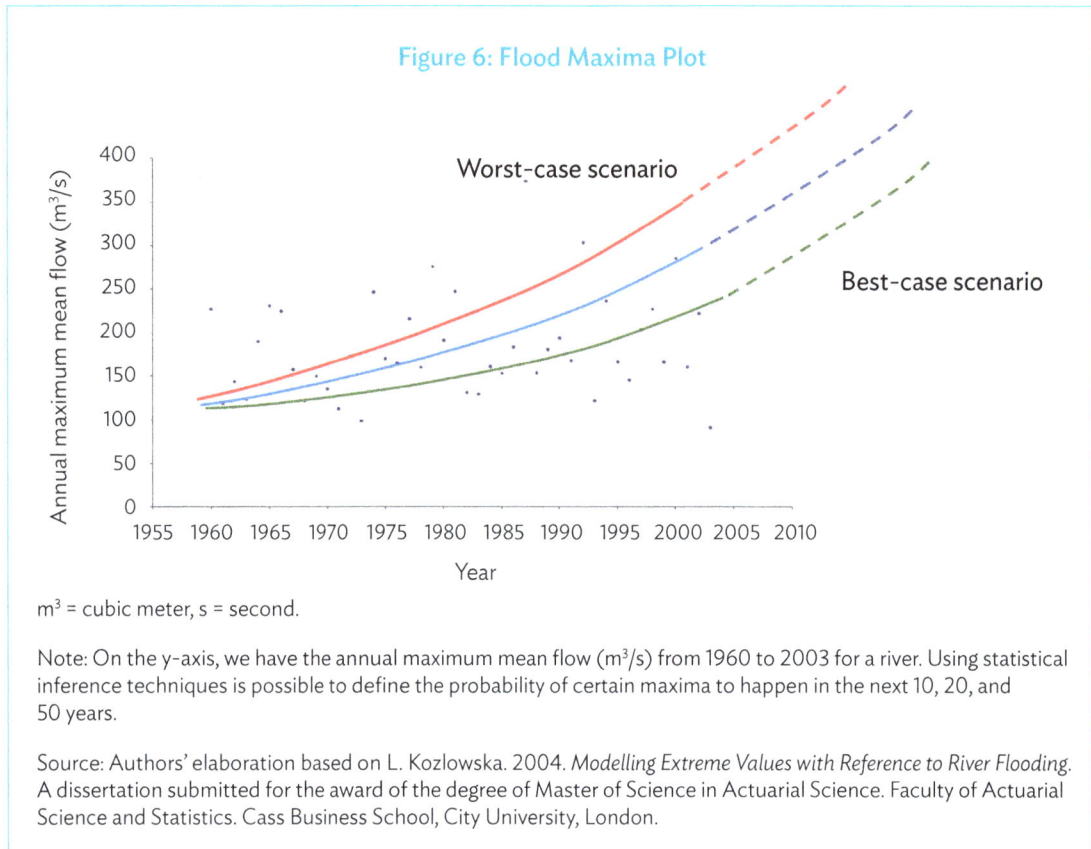

m³ = cubic meter, s = second.

Note: On the y-axis, we have the annual maximum mean flow (m³/s) from 1960 to 2003 for a river. Using statistical inference techniques is possible to define the probability of certain maxima to happen in the next 10, 20, and 50 years.

Source: Authors' elaboration based on L. Kozlowska. 2004. *Modelling Extreme Values with Reference to River Flooding*. A dissertation submitted for the award of the degree of Master of Science in Actuarial Science. Faculty of Actuarial Science and Statistics. Cass Business School, City University, London.

Bridge in city environment. In the event water reached the deck, its wedge-shaped profile would help reduce the pressure exerted by water flowing downstream (photo by Michel Di Tommaso).

E. Floods: Influence of Water Saturation on the Bearing Capacity of Soils

Introduction

In the case of transport infrastructure, particularly for roads, the primary consequence of excessive and rapid saturation during flooding is the softening of foundation materials. These are generally made of a mixture of gravel, sand, silt, and clay. During periods of drought, silt and clay dry, shrink, and crack. When rewetted, they expand. These wet and dry cycles influence the bearing capacity of soils, which is the ability to resist external loads with little deformation. Bearing capacity is a term used to define how well a layer of a pavement material responds to applied stresses exerted by traffic.

Deformability and Stiffness of Unbound Materials

The deformability of soils used to build the foundations and the base-course layers in pavement construction is controlled by a stiffness parameter called resilient modulus (M_r) (Appendix 2). This is the stiffness of soils measured with the assumption that they behave like elastic materials. This assumption is also valid for soils used in pavement construction. The traffic induces load repetitions to a soil in service, bringing the soil grains closer and closer together and maximizing the grain-to-grain contact, which causes the soil to behave like an elastic material. An ideally elastic material is a material where, for any increase in load, there is a proportional increase in deformation and, when load is removed, deformation goes to zero.

The resilient modulus is equivalent to an elastic shear stiffness controlling the amount of shear strain induced by external loads applied to granular and cohesive soils. Because soils have both zero compressive and tensile strength, the only possible way they may deform is through shear (sliding of grains past each other). Hence, maximizing the resilient modulus of foundation materials (also known as subgrades) limits the vertical strain induced by traffic at the interface between subgrade and granular base. Figure 7 demonstrates the various layers making up an asphalt road.

The resilient modulus of soils depends mainly on the following parameters:

(i) degree of saturation (above or below the optimum moisture content);
(ii) climate;
(iii) geotechnical classification (e.g., American Association of State Highway and Transportation Officials, ASTM International); and
(iv) state of stress.

Figure 7: Layers of an Asphalt Road

Source: Authors.

Typical values of resilient modulus of well-compacted subgrade soils at optimum moisture content are generally from 25 to 75 megapascal (MPa), and the higher values are typical in soils with less clay and silt and with more sand and gravel.[19] Experience shows that, when the material is progressively saturated, the value of the modulus may decrease by up to 50%. In fact, when the soil is saturated, the internal grain-to-grain friction is reduced, and it takes little stresses to cause large shear strains because water pressure reduces friction (Appendix 2). This means that wetting reduces the resilient modulus. It takes no more than a few hours to a few days for soils to saturate during extreme rain events, which can reduce dramatically the bearing capacity of a road where flooding occurred, or in an area where intense rain has fallen over a prolonged period.

Granular materials used as bases, subbases courses, and drainage layers have higher resilient moduli than fine-grained soils, with typical values of 200–500 MPa at optimum conditions of compaction and moisture.[20]

---

[19]   In road engineering, the term "optimum water content" refers to the quantity of water allowing maximum compaction, based on Proctor's theory of compaction. The definition of stiffness is the ratio between applied stress and induced deformation. The unit of measure of stress is MPa and the unit of measure of deformation is percentage (%). Hence, the ratio, or stiffness, has a unit of measure of MPa or its multiple, gigapascal (GPa).

[20]   H. Y. Huang. 2003. *Pavement Analysis and Design*. Pearson College Publishing.

Deformability and Stiffness of Bound Materials

Concrete-like materials have compressive, shear, and tensile strengths because the cement holds the grains together within the concrete matrix; hence, when they are loaded, they cannot slip past by each other until the cohesive force of the cement hydrates is overcome.[21] When a concrete pavement is loaded, the layer of concrete is usually receiving tensile stresses at the base. If tensile stresses are large and exceed the tensile strength, the concrete may crack and the pavement could fail and get damaged. Figure 8 illustrates the bending of concrete-like materials due to tensile stress.

To reduce the risk of high shear strains in saturated soils within the subgrade, or in the granular materials within the pavement structure (base course), adaptation measures can be taken to stabilize the soils and/or the granular base courses with a binding agent. When grains are bound by a cement, the mechanical behavior changes and the stabilized material behaves more closely to concrete, providing bending stiffness to resist deflection. Unlike that of the resilient modulus, the degree of saturation of a stabilized material has negligible influence on the value of the bending stiffness, which does not change much from below optimum to above optimum water content.

The bending stiffness of stabilized soils may vary from as low as 0.5 gigapascal (GPa) to as high as 20.0 GPa, depending on the nature of the soil and the amount of cement used. Very fine-grained, sandy, and silty soils generally achieve less bending stiffness than granular ones, such as mixtures of sand and gravel. Typical values for stabilized granular bases made of gravel and sand are between 5.0 and 20.0 GPa, while fine subgrade materials may successfully achieve 0.5 to 5.0 GPa when stabilized with cement. Using the values of resilient modulus for subgrade materials at an average of 50 MPa (or 0.05 GPa), the effect of stabilization is to increase stiffness by many times, meaning that the vertical strains on top of the subgrade are—keeping all the other factors constant—largely reduced (Appendix 2).

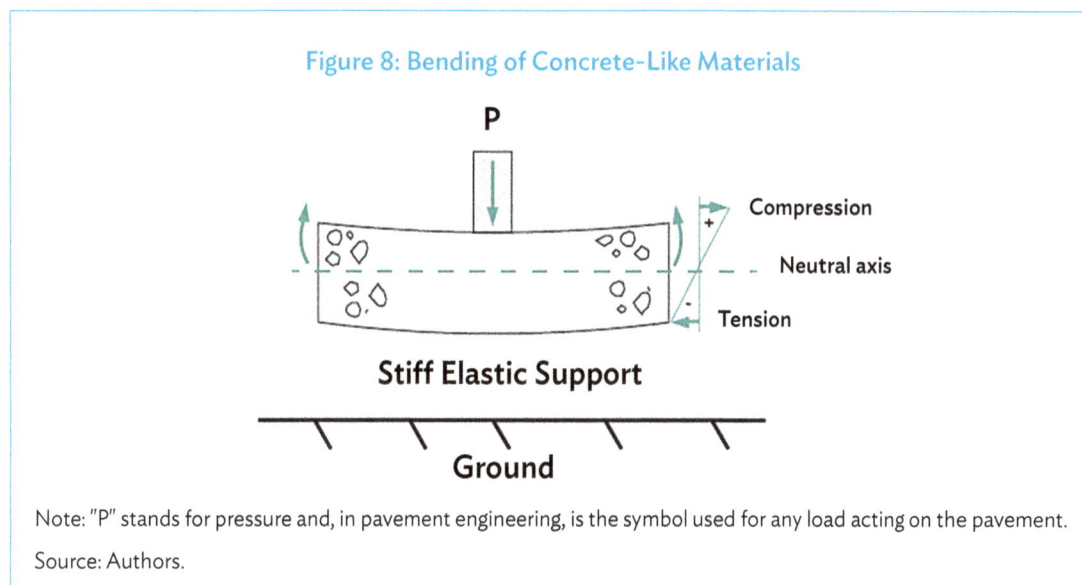

Figure 8: Bending of Concrete-Like Materials

Note: "P" stands for pressure and, in pavement engineering, is the symbol used for any load acting on the pavement.

Source: Authors.

---

21   Concrete-like materials include cement-stabilized granular bases with more than 100–150 kg/m³ of cement.

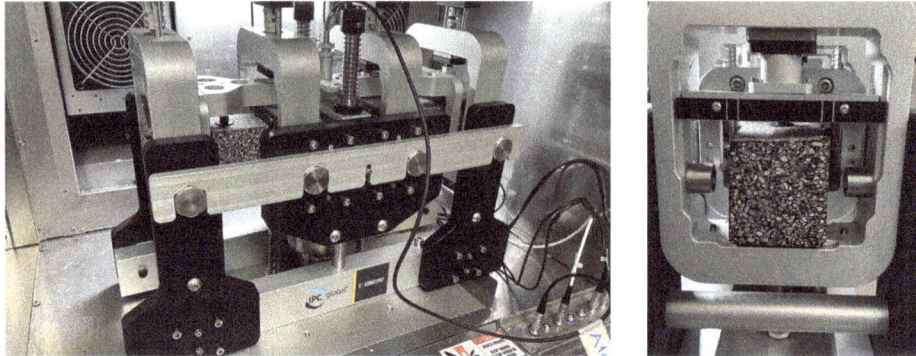

Calculating the bending stiffness of stabilized soil. The photos show the side view (left) and front view (right) setup for measuring the four-point bending stiffness of cement-treated soil specimens (photo by Michel Di Tommaso).

F.   Floods: Influence of Water on Drainage Systems

Introduction

There are two types of drainage systems in the transport infrastructure—surface and subsurface. A surface drainage system collects and diverts stormwater from the road surface and surrounding areas to avoid flooding through culverts, ditches, curbs, and gutters. A subsurface drainage prevents water from entering the pavement from either a slope or from below the road (Figure 9). Drainage systems reduce the rate of water infiltration into the road structure, limiting the softening effect and the loss of resilient modulus.

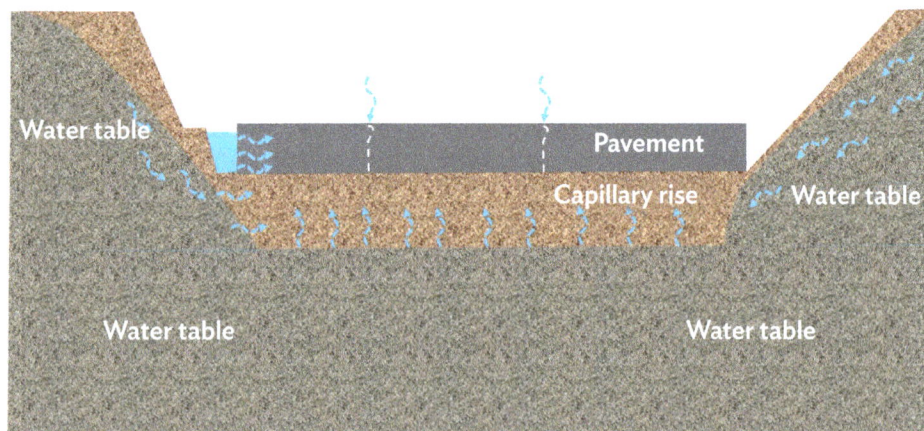

Figure 9: Sources of Water for the Pavement

Note: The sources of water in the pavement are from the side and bottom (water table) and from the top (through cracks and unsealed joints).

Source: Authors.

Surface Protection of Pavement and Subgrade

A key issue related to surface drainage systems is avoiding stagnation of water for a prolonged period. Stagnation of water at the outlet of a culvert or in a ditch causes water to seep through the embankment and through the pavement structure, thus weakening the bearing capacity. Regular maintenance of culverts and ditches is required to avoid blockage by debris transported during rain events.

More frequently than not, drainage systems are underdesigned and poorly maintained, and this is generally true for many regions of the world. Designing an efficient drainage system against the effects of climate change requires a conservative approach in terms of return period of critical surface runoff conditions. Hence, the yearly maxima of surface runoffs should be considered because the intensification of rain events expected in the region will certainly increase over the coming years.

The asphalt and concrete layers used to pave motorways are themselves protecting the underlying materials from wetting caused by surface runoff water. When these layers are cracked or damaged, and/or when concrete pavement joints are no longer sealed, they cause moisture in the base and subbase courses. This indicates that the wearing course of roads either made with asphalt or concrete must be periodically maintained and cracks must be sealed to create an impermeable layer protecting the pavement materials from softening.

The effect of extreme rain events on poorly maintained drainage systems. Ponding water causes soils to soften, and the capillary rise may cause water to rise to the road level above (photo by Michel Di Tommaso).

Well-maintained culvert discharging into a stream. Culverts are important in managing the flow of water and reducing flooding (photo by Michel Di Tommaso).

Protection of Pavement and Subgrade from Lateral Flow and Capillary Rise

Pavements and embankments rest on natural ground that may become saturated during rain events and flooding conditions. If the ground under a pavement structure becomes saturated, water can rise through capillarity in the upper structural layers of the pavement and can reduce the resilient modulus. Also, if a slope becomes saturated, water will seep downhill and may penetrate and saturate the structural layers of a pavement structure. Capillary rise is the vertical height that water can travel above the saturated zone (water table) because of capillary forces. As shown in Table 1, this height is larger for fine soils than for coarse soils.

Well-maintained energy breaker and collection drain. Proper maintenance of drainage systems helps make them sustainable and more effective in protecting transport infrastructure (photo by Michel Di Tommaso).

Table 1: Capillary Rise Values for Different Soils

Material	Particle Size Range (mm)	Capillary Rise (mm)
Fine sand	0.05–0.25	300–1,000
Medium sand	0.25–0.50	150–300
Coarse sand	0.50–2.00	100–150
Well-graded sand	0.25–2.00	150–1,000
Fine gravel	2.00–6.00	20–100
Coarse gravel	6.00–20.00	5–20
One-sized aggregate	>5.00	<5.00

mm = millimeter
Source: Authors.

When pavement structures are built over fine soils that can periodically retain large volumes of rainwater, water moves upward by capillarity and saturates the structural granular layers of the pavement (base course) even in summer, when rain is scarce. When this happens, the bearing capacity of such layers is reduced all year-around and the road may be damaged at a faster rate. Figure 10 shows a typical section of a pavement with a drainage layer to break the capillary rise from underneath.

The effect of a capillary rise. A trench alongside a concrete road after months of drought shows a wet and clay- and silt-rich granular base that is probably saturated by capillary rise (photo by Michel Di Tommaso).

Figure 10: The Influence of a Drainage Layer on Water Movement

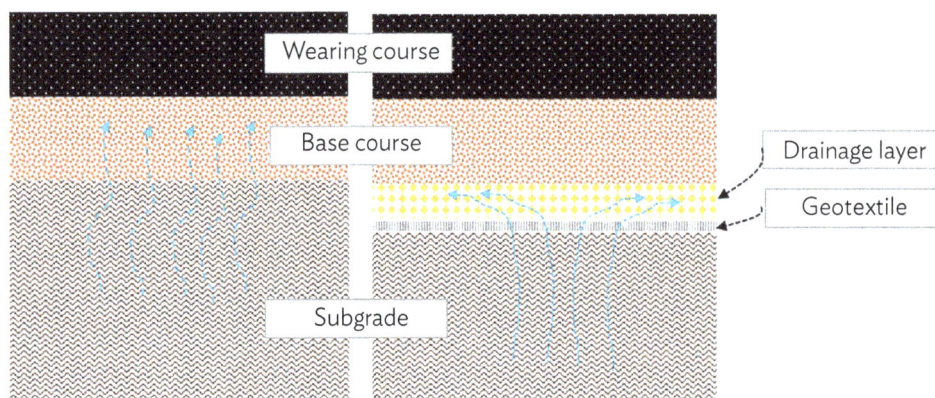

Note: Drainage layers have resilient moduli comparable to granular base courses, thus adding them in the pavement section does not reduce the overall bearing capacity.

Source: Authors.

To control and reduce seepage from the water table, a horizontal drainage layer should be added within the pavement structure to intercept and drain away the capillary rise. Coarser materials have low to negligible capillary rise, therefore drains within the pavement structure are generally made of gravel and coarse sand with little to no fine sand and silt. Geotextile materials are often inserted at the base of the horizontal drainage layer within the pavement structure to avoid the upward seepage of clay and silt particles from the underlying soils, which could reduce the drain efficiency over time. However, geotextiles alone cannot reduce capillary rise.

G.   Floods: Influence of Water on Scouring around Bridge Piers

Bridges are founded on piers that end on pile caps. Together, these group piles allow a bridge to remain anchored to the ground through gravity and friction. Water flowing past a pier will interact with it, and the nature of this interaction will vary depending on the pier's geometry and the number of piers involved. The depth of scour, keeping all the parameters constant, depends essentially on flow depth and on the width of pier or groups of piers or abutment.

Local scour under bridge foundations can be classified as live bed or clear water conditions. In clear water scour, there is no sediment available from upstream when the scour hole is forming downstream. When water is transporting a bed load of sediment, there is a supply of new sediment from upstream. Hence, a scour hole is likely to increase in size when clear water is involved in the process. The shape of piers influences the degree of scour, with cylindrical and/or elliptical shapes considered as the baseline for design.

The depth of scouring can be predicted by applying empirical relationships that take into account input parameters such as velocity of flow, nature of sediment load (suspended, bottom, or clear water), shape of the channel, and shape of the pier and/or pile cap.

Fast-moving, turbulent, clean water has the highest scouring effect because no sediment is transported and deposited in the scour hole during the process.[22] Designers should consider the risk of exceeding a certain scour depth when a given geometry of the piers and/or pile caps system is used with worst-case scenarios of velocity of water and sediment load. If the predicted scour depth is excessive and could expose the top of the piles and hinder structural safety, designers should recalculate by changing the geometrical factors of the piers and/or pile cap system to make the anticipated depth of scour sustainable by the structure.

From investigations on the scouring effects on bridge piers, it is evident that local scour countermeasures are required primarily upstream of a bridge pier, where the stronger horseshoe vortex forms, as well as downstream of the bridge pier where another, yet weaker, wake vortex system develops (Figure 11).

Figure 11: Flow Lines of Fast-Moving Water around a Pier

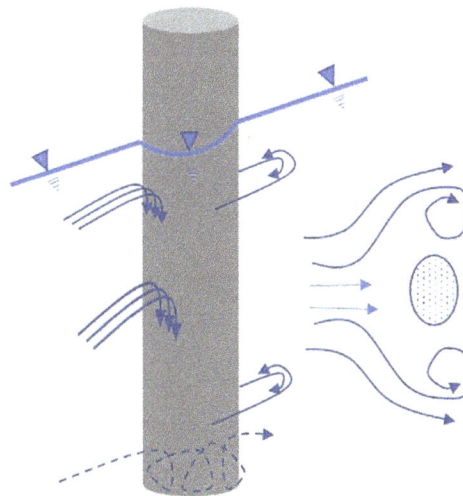

Note: Critical areas for scouring are upstream and downstream of the pier (the flow lines describing the motion of water particles are moving from left to right). It can be noticed that flow lines become helicoidal and/or spiral immediately upstream and downstream of the pier. The combined effect—in three dimensions of rotation both parallel and perpendicular to the riverbed—causes a hole to form downstream of the pier because of the scouring effect.

Source: Authors.

---

22   It is beyond the scope of this manual to illustrate the design process defining the depth of scour. However, to know more about this topic, see, for instance, the work of W. B. Melville. 1997. Pier and Abutment Scour: Integrated Approach. *Journal of Hydraulic Engineering.* 123 (2). pp. 125–136.

H. Floods: Water as a Driving Agent of Gravity Flows

Introduction

Intense rain causes flooding in flat areas and may also mobilize large masses of soils resting on slopes that are both natural and artificial, such as embankments. Since the pressures exerted by moving mixtures of soils and water are very large, the impact of these phenomena on the transport infrastructure should be understood and considered. In view of climate change projections for Central Asia, the intensity and frequency of such events will most likely increase in the years to come.

Mamadjanova et al. (2018) reported data on the damaging effects of mudflows in Uzbekistan over 2005 to 2014.[23] Such events took a toll on people, transport infrastructure, and the economy (Table 2).

Table 2: Damage by Gravity Flows in Uzbekistan, 2005–2014

| Year | Deaths | Household or Property Damage | Livestock Counts | Infrastructure Damage | | | | |
				Motorways (km)	Bridges (km)	Dams	Schools	Others
2005	...	860	...	...	1	...	...	2
2006	7	175	...	...	...	...	...	2
2007	...	8	1	6	15	7	...	3
2008	7	413	1	0.35	5	...	...	49
2009	8	498	80	...	14	5	2	...
2010	8	41	...	...	6	...	2	7
2011	2	94	50	0.5	...	1	...	52
2012	5	773	3	2.7	25	6	1	55
2013	1	31	...	0.012	2	6	...	3
2014	...	...	...	...	...	...	...	4
TOTAL	38	2,893	135	10	68	25	5	177

... = data not available, km = kilometer.

Source: Reworked from G. Mamadjanova et al. 2018. Statistical Characteristics of Mudflows in the Piedmont Areas of Uzbekistan and the Role of Synoptic Processes for their Formation. *Natural Hazards and Earth System Sciences.* 18 (11). pp. 2893–2919.

---

[23]   G. Mamadjanova et al. 2018. Statistical Characteristics of Mudflows in the Piedmont Areas of Uzbekistan and the Role of Synoptic Processes for their Formation. *Natural Hazards and Earth System Sciences.* 18 (11). pp. 2893–2919.

Gravity Flows: Basic Concepts

Gravity flows are phenomena whereby a mass of soil and/or rock is mobilized and flows downward from a position of high potential energy to a position of low potential energy through the creation of kinetic energy and heat on its path. The kinetic energy is the potentially disruptive force of concern in terms of structural failure or damage of any infrastructure that would be on the path.

Triggering events for these flows are related to water saturation, earthquakes, soil depletion due to fire, and explosive volcanic eruptions causing lahars,[24] among others. Mudflows and debris flows are expressions of essentially the same dynamics, while snow avalanches, another threatening agent for the infrastructure in mountain areas, are treated separately. The transition from mudflows to debris flows depends essentially on cohesion, grain size, and volume fraction of water and solid.

Genuine mudflows require the presence of abundant cohesive materials (silt, clay, organic soils, etc.) and water. They tend to move rapidly. Debris flows also require abundant granular materials (sand, gravel, cobbles, boulders) and water. They also tend to move rapidly, given the right conditions. Between these two extremes are mixed flows involving both cohesive and granular materials, making the distinction rather arbitrary. Figure 12 presents the different classification of gravity flows.

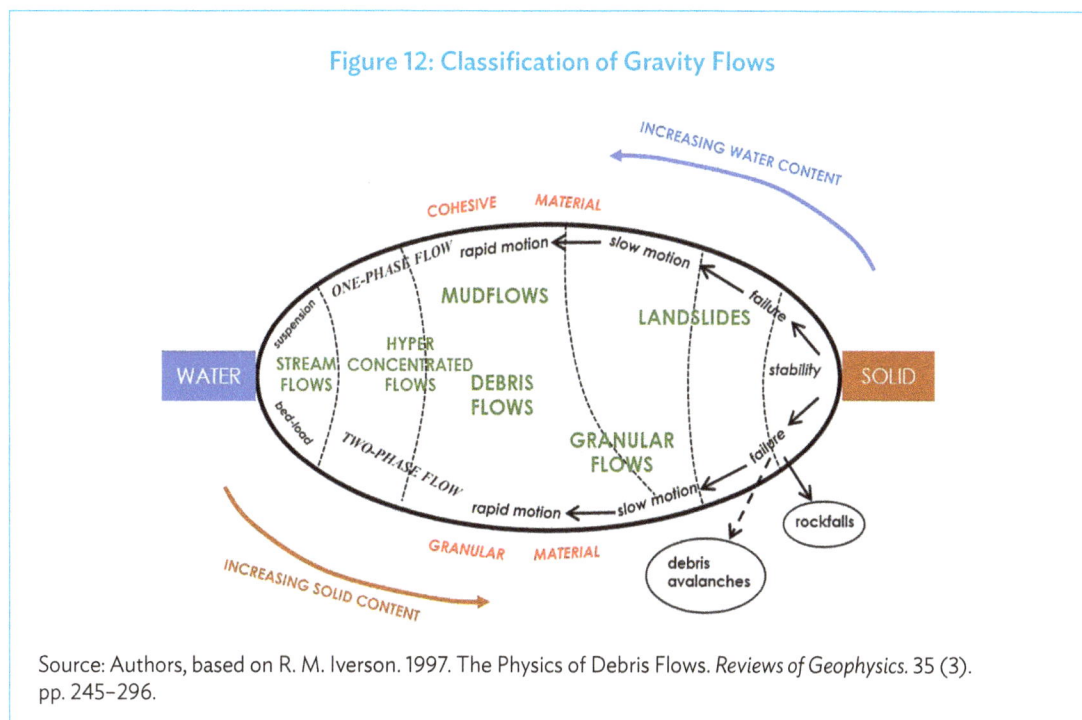

Figure 12: Classification of Gravity Flows

Source: Authors, based on R. M. Iverson. 1997. The Physics of Debris Flows. *Reviews of Geophysics.* 35 (3). pp. 245–296.

---

[24] Lahar is an Indonesian term referring to flows of mud, ash, and rock caused by the deposition of unconsolidated layers on the flanks of a volcano following an explosive eruption. During the rainy season, these layers may be mobilized and may flow downstream with extremely disruptive forces.

When dry or at its natural moisture content, soil is generally at rest even if the slope is relatively steep. Upon saturation by rain, internal friction is suddenly reduced and shear stresses pulling the mass of soil downward increase, thus the soil can flow downhill (Appendix 3).

R. M. Iverson (1997)[25] reports examples of massive debris flows induced by volcanic eruptions, such as those at the United States' Mount St. Helens in 1980, travelling a distance of 45 km for an estimated volume of 10^7 m^3, versus a 10^3 m^3 volume landslide travelling only about 0.2 km.

Soil grains during flow can either be transported in suspension within the fluid fraction because of viscous forces (this is true for clay and silt that create slurries), or else they can be transported in suspension only if continually impacting on other grains (dispersive pressure). During the interaction of grains, dispersive pressure acts as the force normal to the plane of shearing, which tends to expand or disperse the grains in this direction.

High dispersive pressure in debris flows is favored by hard and nonweathered rock fragments, allowing them to roll and bounce elastically—keeping all the other factors constant—and more efficiently than in debris flows made of weathered and/or weak rocks.

According to Iverson, the key properties of gravity flows are the following: unit weight, grain size, friction angle of the soil grains, hardness of the rock fragments, and volume fraction of solids (footnote 25). The unit weight of debris flows exceeds 2,500 kg per m^3, which is higher than the density of fresh concrete (2,400 kg per m^3) and indicates that, when moving down a slope, these flows can exert enormous pressures on anything they find on their path. Slope failures along transport infrastructure can be considered as a very special case of gravity flows, and they are governed by the similar geotechnical and geometric parameters explained in Appendix 3.

I. Temperature-Related Stressors on Transport Infrastructure

Introduction

The projected scenarios for temperature increases in the Central Asian region up to year 2085 indicate that the duration of heat waves and the maximum temperature recorded during heat waves will likely increase. The average temperature in the region may rise from the actual values to about 2°C above pre-industrial levels in best-case scenarios, to more than 6°C above pre-industrial levels in worst-case scenarios. In engineering, conservative assumptions are factored into the calculations.

Taking an average assumption of a 4°C rise, summers would be hotter and winters warmer. Many of the inhabited areas of the region experience continental climate with long, dry, and hot summers and irregularly cold winters. Air temperatures in the desert plains currently reach up to 45°C; while, in the

---

[25]   R. M. Iverson. 1997. The Physics of Debris Flows. *Reviews of Geophysics*. 35 (3). pp. 245–296.

Slope failure along a major highway in Central Asia. The slope was cut on a plane inclined at an angle higher than the angle of internal friction of the soil. The slope is close to an angle of 45°, while the nature of the soil exposed, rich in silt and clay, would relate to an angle of friction of not more than 35°. The height of the slope is also excessive for the steep angle chosen (photo by Michel Di Tommaso).

Using berms to avert slope failure. This slope is more than 40 meters high, with only one berm to break down the pulling forces, which is not enough to control sliding. Typically, for every 10–12 meters of height, horizontal berms should be constructed to avoid slope failure (photo by Michel Di Tommaso).

winter, temperatures may drop to as low as −25°C. Even in areas where maximum temperature during the summer does not exceed, on average, 35°C, temperatures will probably shift toward higher average values in the future.

Therefore, areas that currently have milder temperatures may become areas where extreme heat waves hit harder in the future. This means that the road transport infrastructure will be subjected to both very high temperature differences between the summer and the winter (possibly up to a 75°C to 80°C difference between the two seasons in extreme cases). It could also face extreme summer temperature for prolonged periods.

Effect of Temperature on Asphalt Pavement

A flexible pavement is made of asphalt concrete, a material composed of aggregates and a polymeric binder or bitumen. Like all polymers, bitumen is temperature-sensitive and its hardness increases at low temperature and decreases at high temperature. At room temperature, it behaves like a stiff solid when loaded with a dynamic (pulsive) load, and it behaves like a fluid when the applied load is static. This means that bitumen becomes "softer" when temperature increases and/or when traffic moves slowly, while it gets stiffer when temperature is low and/or traffic is fast. During the exceptionally intense heat wave in southwestern Canada in the summer of 2020, where air temperatures of up to 50°C were reached, there were reports of "molten" roads where asphalt had changed from solid to a viscous fluid. As a first approximation, the surface of asphalt is about 10°C warmer than ambient temperature under direct insolation, but the temperature will rapidly drop with depth, and it will stabilize to ambient temperature a few centimeters below the surface. Hence, if air temperature is, for example, above 40°C with clear skies, the top few centimeters of the asphalt may reach a temperature of more than 50°C.

Effects of extreme temperatures on asphalt. Rutting forms on the asphalt pavement in the summer (left photo), while cracking forms in the winter (right photo) (photos by Michel Di Tommaso).

This has important consequences. When bitumen has heated up, the vertical stresses exerted by traffic on the top asphalt layer pavement become concentrated; however, when bitumen has cooled down, the stresses in the pavement are well spread. In the first case, shear failure of the asphalt layer is possible, and rutting starts to form.

Although cold temperature increases the stiffness of asphalt, the downside of this is that the asphalt becomes more brittle and therefore prone to cracking. A study of the influence of temperature and the frequency of pulse loads on asphalt takes place in a laboratory using stiffness tests. In these tests, a sample of asphalt shaped like a cylinder or a beam is subjected to an impulsive tensile or flexural load in a temperature-controlled environment. Varying the frequency of the loading pulse (measured in hertz) and the environment temperature can determine the stiffness of asphalt as follows:

$$tensile\ or\ bending\ stiffness = \frac{tensile\ or\ bending\ load}{recorded\ elastic\ displacement}$$

A test setup to measure the stiffness modulus of asphalt in the laboratory is shown in the photos below.

Laboratory setup to test stiffness of asphalt cores. Load is pulsive and applied vertically, while displacement transducers record lateral strain. The ratio between applied stress and induced strain is the stiffness modulus of asphalt (photo by Michel Di Tommaso).

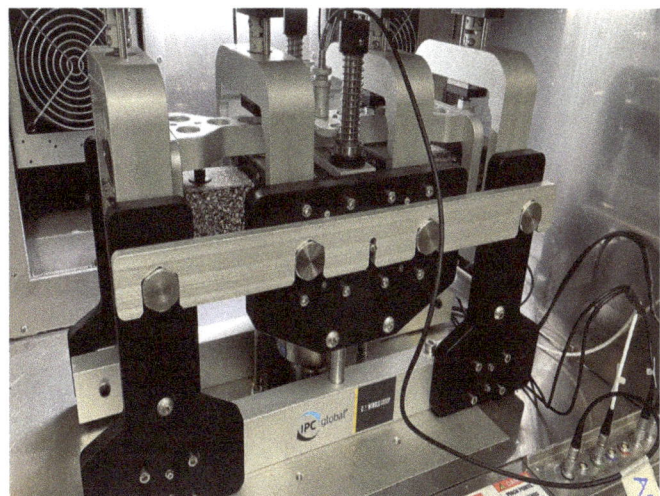

Laboratory setup to test stiffness of asphalt beams. Load is pulsive and applied vertically, while displacement transducers record bending strain. The ratio between applied and induced strain is the stiffness modulus of asphalt in flexion (photo by Michel Di Tommaso).

These tests indicate systematically the peculiar behavior of asphalt concrete. At low frequency of load, the stiffness decreases, while at high frequency of load, the stiffness increases. Low frequency loads (<1 hertz) simulate static traffic loads, while high frequency loads (>10 hertz) simulate fast-moving traffic.[26] This means that, to provide realistic values of asphalt stiffness for the design of roads, it is important to define the boundary conditions of the stiffness test, since the same asphalt concrete mix may display varying levels of stiffness depending on the temperature and frequency of the pulse load applied. Also, regardless of the type of asphalt concrete, it is observed in general that, above 40°C, the stiffness modulus drops by 10 or 20 times with respect to the stiffness values measured at, for example, 10°C to 15°C.

A well-designed binder course may have stiffness modulus of up to 6.0–8.0 GPa at 10°C and drop to less than 600 MPa at 40°C. Usually, in extremely hot weather with air temperature of up to 45°C and with high solar radiation, asphalt concrete heats up for many centimeters below the surface, meaning that the upper layers of the asphalt pavement (wearing course and binder course in conditions of extreme heat) could reach values of temperature causing the stiffness to collapse. When this happens and the road is trafficked, specially by heavy and slow-moving heavy vehicles, rutting forms (Figure 13).

Likewise, in the winter, the temperature may be close to or below freezing in the same areas experiencing hot summers (this is typical of continental climate) and the stiffness modulus would increase and become maximum. With high stiffness, even the smallest contraction caused by thermal shrinkage of the asphalt from summer to winter will translate into high tensile stresses, which can easily exceed the very low tensile strength of asphalt concrete and cause it to crack because of thermal fatigue.[27] In an area where the temperature difference from summer to winter is as high as 60°C, asphalt will contract twice as much in passing from summer to winter with respect to an area where the difference is only 30°C. The phenomenon of thermal cracking for concrete materials is entirely applicable to cold asphalt too, behaving, for any practical purpose, like a brittle–elastic material at low temperature.

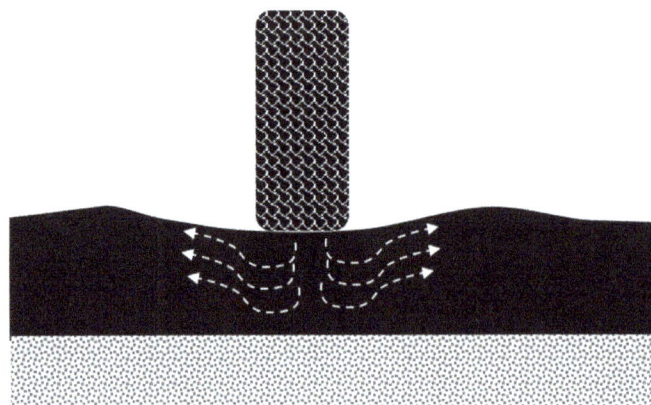

Figure 13: The Mechanism of Rutting of Asphalt

Source: Authors.

---

[26]   Static traffic is like trucks queuing on a carriageway, or a fleet of airplanes resting on an airport taxiway.
[27]   Repeated cycles of summer expansion and winter contraction cause thermal fatigue.

To adapt against the thermal effects on asphalt due to increase of air temperature and temperature extremes, bitumen should be engineered to make it less temperature-sensitive and more resistant to rutting in general. Thus, choosing a concrete pavement instead of an asphalt one may be considered a viable adaptation against climate change.

J.   Effects of Temperature on Concrete Pavement

Concrete, unlike asphalt, does not change stiffness with changing temperature (within the ranges expected from climate change) and with changing pulse load intensity. This is because the crystalline nature of the cement does not make it temperature-sensitive. But concrete can expand and contract more than asphalt because the coefficient of thermal expansion of concrete is at least 10 times larger.

The coefficient of thermal expansion of concrete (α_c) defines the amount of expansion when concrete is heated-up from a lower to a higher temperature, as well as the amount of contraction when concrete is cooled-down from a higher to a lower temperature. This coefficient depends mainly on the nature of the aggregates, and it may be assumed, for design purposes, to be 12 $\mu\varepsilon$ per °C (or 0.0012%).

Hence, for every increase or decrease of temperature of 1°C, a unit length of concrete expands or contracts by 0.0012% of the original length. The total variation in length (or strain) is given by

$$\varepsilon = \Delta T \cdot \alpha_c$$

where: ΔT = temperature difference (°C)

Buckled concrete pavement in Central Asia. The thermal expansion of concrete as a result of changing temperature is more evident than that of asphalt (photo by Michel Di Tommaso).

Concrete pavements are built when the air is neither too cold nor too hot (usually between 5°C and 35°C) because fresh concrete suffers from both heat and cold. The Worked Example in the box below illustrates how the difference in temperature from the moment the pavement is formed to the maximum temperature in service for the pavement dictates the spacing of expansion joints.

Usually, concrete pavements have expansion joints located at equally spaced distances to ensure that the concrete pavement between two of these joints can expand without buckling.

Worked Example

- The expansion ΔL experienced by a slab of concrete of length L and coefficient of thermal expansion α_c under a thermal difference ΔT (difference between T2 [maximum average temperature in service] and T1 [average temperature when concrete was placed]) is given by $\Delta L = R \cdot L \cdot \alpha_c \cdot \Delta T$.
- R is a coefficient of restraint varying from R = 1 (no expansion, full restraint) to R = 0 (full expansion, no restraint).
- For concrete resting over a subbase, only the upper portion of the thickness will be free to expand, while the bottom will be fully restrained.
- We assume R = 0.25 to account for only a partial expansion of the concrete slab, which is fully bonded at the base and free to move only toward the top.
- Granite rocks are commonly used in Uzbekistan to pave concrete, with granite having
$$\alpha_c = \frac{0.0012\%\cdot}{°C}$$
- Expansion joints should have $\Delta L \leq 3$ centimeters (cm) to ensure load transfer efficiency across the joint, which would be lost if the joint has a wider opening when built.
- We can now see the maximum distance between two expansion joints depending on the value of ΔT.
- The calculations reported below shows that slabs cast in the winter (25-meter spacing for T1=5°C) require much more frequent expansion joints than slabs cast in the summer (200-meter spacing for T1= 35°C) when T2= 40°C is the average maximum summer temperature of the pavement.

L	25'000 cm	40'000 cm	200'000 cm	ANY cm	Unit
ΔT	0.0012%	0.0012%	0.0012%	0.0012%	$\mu\varepsilon/°C$
$T1$	5	15	35	40	°C
$T2$	40	40	40	40	°C
ΔT	35	25	5	0	°C
R	0.25	0.25	0.25	0.25	
ΔL	3	3	3	0	cm

Source: Authors.

This is also an issue for bridge abutments, where poorly designed expansion joints may cause the expanding concrete pavement to exert loads of several tons on bridge abutments causing compressive and shear failures. To avoid damage to the concrete road and abutments because of increased climate-change-related expansion, concrete pavement must be designed with an adaptation strategy in mind that aims to control the risk of blowups (Figure 14).

Figure 14: The Mechanism of Blowups in Concrete Pavements

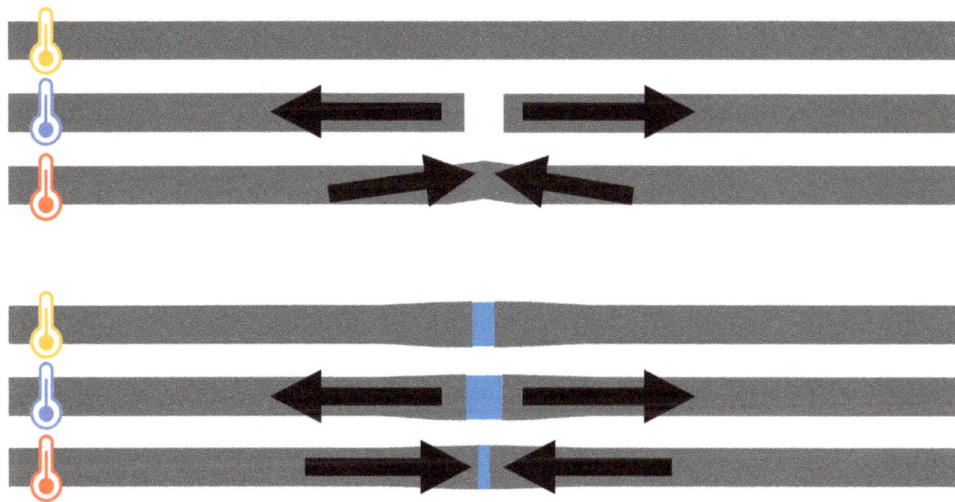

Above: Concrete is cast at mild temperature (orange) without an expansion joint. It will contract in the winter and expand in the summer beyond the original length causing blow-ups.

Below: Concrete is cast with an expansion joint. Upon expanding in the summer, it will have room to accommodate the extra length, and the two sections will not compress. For this reason, the joint is filled with a compressible material chosen to have sufficient width to compress and extend following the joint's movements.

Source: Authors.

K. Snow Piling and Snowcaps Melt Impact on Transport Infrastructure

Introduction

The intensification of extreme events associated with climate change may also cause the sudden accumulation of snowcaps in winter. Rapid accumulation of wet snow on slopes, when the air temperature is close to or just above 0°C, can trigger disruptive snow avalanches causing damage to transport infrastructure in mountain regions. As with sliding soils, when snow is wet, it weighs more and, therefore, the tangential pulling stress parallel to the slope exceeds the vertical stress holding the mass of snow pressed onto the slope.

Rapid melting of glaciers can cause tension relief stresses in rock masses that can break, topple, and fall, thus hindering the safety of nearby roads and railways. The melting of permanently frozen ground (permafrost) is also causing rock masses to destabilize because of gravity.

Snow Piling

The rapid movement of masses of snow or ice down the slopes can pose a serious hazard in mountain areas. For example, snow avalanches, particularly when they contain large amounts of debris, can damage buildings and motorways and lead to loss of life. Two types of avalanches are recognized: the dry snow and the wet snow. Avalanche location can often be predicted from historical evidence relating to previous events combined with topographical data. Consequently, hazard maps of avalanche-prone areas should be produced as part of adaptation measures and should be used when designing resilient road infrastructure. With atmospheric moisture increasing because of the increase in average air temperature and higher evaporation rates from oceans induced by global warming, extreme rain events seen during warmer seasons can turn into extreme wet snow piling events in the winter in mountain areas.

Higher average winter temperatures can lead to massive accumulations of wet snow on slopes, which are unprotected by vegetation that may have been lost because of summer fires or anthropic activities. As discussed earlier, how slopes made of soils become stable, the formation of snow avalanches involves similar mechanics and dynamics, where a pulling shear force parallel to the slope is resisted by the friction and cohesive forces that the snow mobilizes within its mass and with the ground. The pulling stress τ is a function of the angle of the slope, the thickness H of the snow, and the unit weight γ of snow. If snow begins to melt and water is trapped at the interface between the snow and the ground, friction is reduced, and sliding can start (Figure 15).

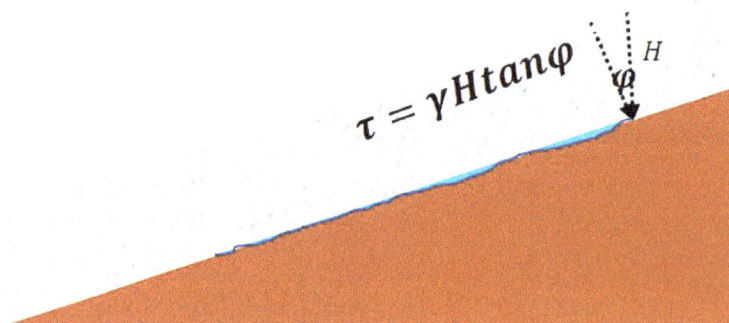

Figure 15: A Slab of Snow Resting on an Inclined Slope

$$\tau = \gamma H \tan\varphi$$

Note: When water from melted snow (blue line in the sketch) is present between snow and ground, it will reduce friction and cause slip.

Source: Authors.

Averting snow avalanches. Naked and steep slopes are protected against avalanches with structural fencing (photo by David Radomysler, pexels.com).

Snow avalanches are made of semicompacted snow that has a unit weight of about 0.4–0.5 tons per m³, which is about five times smaller than the density of a debris flow. Therefore, lower destructive forces are expected to be involved, since the pressure exerted on obstacles is proportional to the unit weight of the mixtures. However, the velocity at which snow avalanches move is generally much higher than debris flows, which makes the impact on infrastructure as damaging and hazardous as other types of denser, yet slower, gravity flows. As discussed in the next section, when snow is expected to accumulate on naked and very steep slopes, it is extremely important to increase the friction between the snow and the ground by adding protective structures, such as the structural fencing shown in the photo above. In other cases, it is safer and more economical to protect roads and railways with artificial box tunnels.

L.   Rockfalls Caused by Tension Relief and Permafrost Melt

The Intergovernmental Panel on Climate Change (IPCC) has published guidelines on the impact of glacial and permafrost melt on human activity and built infrastructure.[28] The piling of glaciers during the last ice age has caused rock masses at the bottom and sides of glacial valleys to remain in a state of compression for tens of thousands of years under the weight of tens or hundreds of meters of compacted ice exerting several MPa of pressure.[29] When this pressure is suddenly removed as ice melts,[30] rock masses near the surface have an elastic rebound, which may cause large-scale tension

---

[28]   The Intergovernmental Panel on Climate Change.
[29]   1 MPa = 10 kg per cubic centimeter
[30]   Geologically speaking, a sudden event takes place within a few decades or centuries.

relief cracks parallel to the profile of the valley. This can cause the sides of the valley to become unstable and slide downstream. After the glacier has melted, the dominating state of stress close to the surface, once covered by ice, is no longer compressive but tensile. This causes rockfalls, rock avalanches, and debris flows to happen more frequently, involving larger volumes of debris.

Besides tension relief, permafrost melt is an issue for the stability of fissured rock masses. Permafrost, by definition, is ground that can remain frozen for at least two consecutive years. Permafrost is related to the average annual temperature of a region and can be preserved only if water is trapped in the ground and within the cracks of rock masses as ice, irrespective of the season. If water penetrates the cracks in liquid form during summer because the ground is not frozen, it will expand in winter when it turns into ice, and some degree of internal pressure will build up within the cracks. If this is a cyclic event (water in summer and expansive ice in winter), the rock mass may become destabilized with time and slide and/or topple depending on the prevailing direction of the vector of gravity.

If the presence of moisture and the periodic change of state from water to ice are the known triggering effects for sliding of rock slopes, the way the rock mass is cracked or layered dictates how transport infrastructure may be affected. Cracks, joints, and strata in rocks should be imagined as three-dimension planes of weakness with a given orientation in space. These features break the homogeneity of rock materials and introduce anisotropic behavior, whereby some directions are more prone to sliding (Appendix 4).

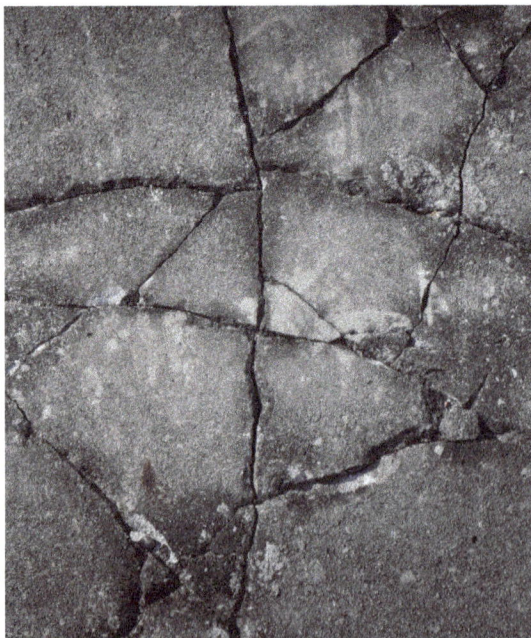

A heavily jointed rock face. This shows different joint planes with crossing orientation, dividing the rock mass into blocks of varying shapes and volume (photo by Tristan Tan, Shutterstock.com).

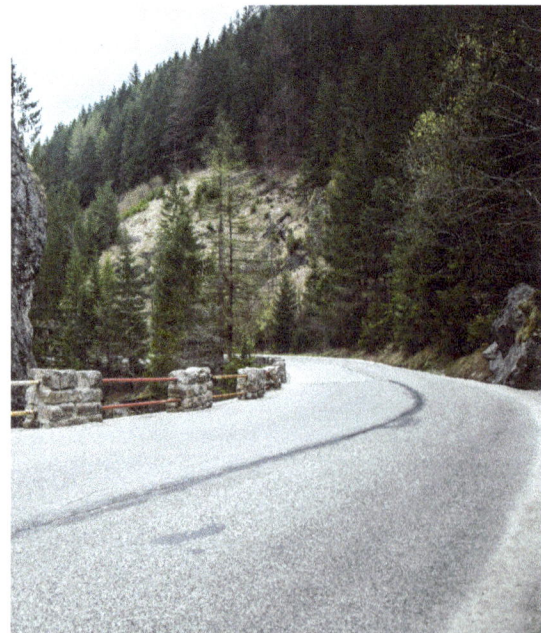

Unfavorable condition for sliding. The rock, a metamorphic rock, is foliated but the general orientation of each plane separating the strata is such that the gravity vector holds the strata together and the whole mass is in compression, thereby largely increasing the shear strength required to mobilize it (photo by Romiya, Shutterstock.com).

M. Effects of Fires on Soil Stability

The consequences of climate change can include prolonged periods of drought and dry spells. Sustained wind conditions and low humidity, associated with very high ambient temperature, can trigger widespread forest fires. Besides adding more CO_2 into the atmosphere, slopes become unstable when the roots of trees and shrubs are burned as their ability to stabilize soils and reduce the potential for sliding of the colluvium[31] during rain events is diminished.

If an area has been affected by fire, and if slope slides during subsequent rain events pose a high risk for the transport infrastructure, structural fencing and the planting of new trees can help the affected slopes regain stability.

A typical example of vulnerability of colluvium soil after fires may be found in California, where steep slopes exposed to both wildfires in summer and extreme rain events in spring and/or winter may generate powerful and destructive debris and mudflows.

Unstable slope triggering the sliding of colluvium. Colluvium is the upper portion of the slope close to surface where roots stabilize the soil (photo by Michel Di Tommaso).

Naked slope from forest fire. With most of the vegetation consumed by fire, the exposed soil is highly prone to landslides and mudflows (photo by Elena Sherengovskaya, Shutterstock.com).

---

[31] Colluvium is the residual soil formed by the chemical and physical weathering of the bedrock that becomes an organic rich soil.

Adaptation to Climate Change to Increase Resilience of Transport Infrastructure

4

A. Introduction

This chapter deals with adaptation strategies required to cope with climate change stressors. For each climate-related stressor previously discussed, adaptation measures are provided based on state-of-the-art solutions available from the engineering community.

B. Flood Adaptation for Transport Infrastructure: Maximizing Resilience

Once the likelihood and the intensity of an extreme flooding event for a region or area associated with a transport infrastructure project have been determined and the site reconnaissance and data analysis carried out, designers should focus on maximizing the resilience of the infrastructure that may be affected.

Resilience must be maximized—having in mind what level of risk is taken—should the threshold values for which the structure is designed be exceeded. Resilience-oriented solutions for protecting transport infrastructure against flood damage may include

(i)   elevating the deck of bridges to levels that are safely above the highest flow recorded (if historic data is available);
(ii) using cementitious stabilizers for soils used in pavement construction;
(iii) increasing strength and resistance to scour and abrasion of concrete structures and foundations exposed to moving water in rivers;
(iv) incorporating additional capacity in drainage design; and
(v)   using reinforced embankments, slopes, and larger spans; and removing piers from the active floodplains.

C. Flood Adaptation for Transport Infrastructure: Stabilizing the Soil

Introduction

Soil stabilization is the process allowing soils to achieve properties that are specifically engineered. Soils can be either granular (such as sand and gravel) or cohesive (such as silt and clay). Granular soils have no internal cohesion, while fine-grained soils have internal cohesion (Appendix 3). Cohesive materials may be present in the foundations of a road or a railway embankment resting on a natural ground, which is generally a mixture of clay, silt, and sand. Granular materials are used for bases, subbases, and drainage layers, which are the essential pavement structural layers. In this manual, when dealing with the engineering aspects of pavements, the term "soil" is used in a broad sense to indicate the clay, silt, sand, and gravel (natural or processed) used to build the pavement section from the natural ground (or subgrade) to the upper asphalt or concrete layer (a typical pavement structure for concrete and asphalt is presented in Figure 16).

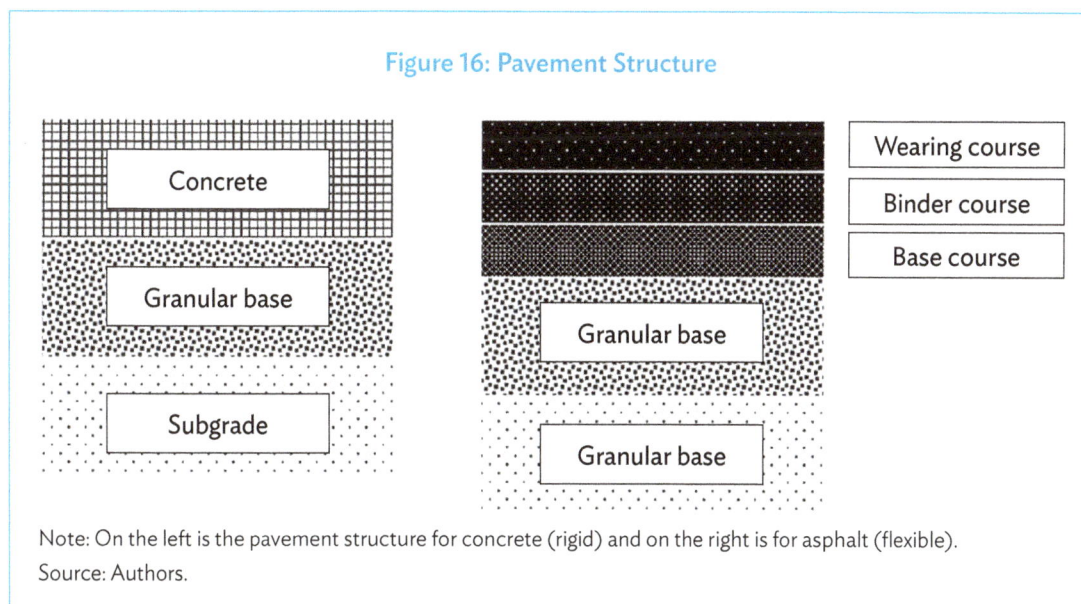

Figure 16: Pavement Structure

Note: On the left is the pavement structure for concrete (rigid) and on the right is for asphalt (flexible).
Source: Authors.

Engineering Modification of Soils

Subgrade and base courses are exposed to saturation because of capillary rise from below and because of flooding of the pavement structure during rain events from the sides and above. The way soils respond to rapid saturation during and after an extreme rain event is an important parameter to understand the causes of the loss of bearing capacity and to prevent such loss.

The stabilization of pavement materials has two scopes:

(i)   providing an engineering improvement to locally available marginal soils when the importation of high-quality materials might be impractical, uneconomical, or restrained by CFP policies; and

(ii)   reducing the material susceptibility to moisture, thereby decreasing deformability and increasing stiffness.

Mitigation measures against climate change should include reducing the CFP of a pavement construction. Since transport has a huge impact on the CFP of construction activities, rather than sourcing high-quality granular base materials from distant quarries, locally available (and lesser quality) materials may be used, provided that an engineering modification is made. The photo on the right shows a sample of an engineering modification.

When soils for road construction are stabilized, the stiffness is increased and they store less permanent deformation. Hence, stabilizing a soil is equivalent

Engineering modification of pavement materials. A cement-stabilized base (white) is used under the asphalt binder and wearing courses to improve the pavement structure (photo by Michel Di Tommaso).

Pulvimixer in action. The subgrade soil for a railway infrastructure is stabilized with lime and cement (photo by Michel Di Tommaso).

Stationary cement-mixing plant. A stationary plant produces cement-treated base materials for road construction (photo by Michel Di Tommaso).

to increasing the bending stiffness and reducing the magnitude of stresses at any depth of the pavement section. The process of stabilization is simple. It consists of either blending the material in situ with a specific binder through a process called in situ mixing, or producing it at a stationary or mobile plant and delivering it to the site using tipper trucks.

In the in situ mixing process, the binder is mechanically spread on the surface of the soil to be stabilized in a specific and controlled amount, then a special machine (sometimes called pulvimixer) follows, which mixes the soil and binder together with water to promote optimum densification. Finally, rollers will compact the layer to a maximum density. Typical stabilized layers for pavements have thickness ranging from a maximum of 0.35 meters to a minimum of 0.15 meters, depending on the type of loads and traffic volumes.

A stationary mixer for producing stabilized granular fills is shown in the photo above. Here, the mixed material (binder plus water plus soil) is discharged into tipper trucks and is transported to the site. From a CFP standpoint, in situ cold mixing has less impact than remote blending. Remote blending requires excavation and hauling of materials, energy consumption at the batching plant, transport, spreading, and compaction, while cold mixing removes hauling and transport, as well as the energy required to batch the stabilized granular material (Appendix 5).

Engineering Modification of Soils in Rural Roads

Soil stabilization is a useful tool for improving the resilience of rural roads, which are generally unpaved and/or lightly paved because of light traffic. Sometimes, they are paved with a thin asphalt layer, which is used mainly to reduce the penetration of water from the top.

Rural road construction. Rural roads are usually either unpaved or lightly paved because of light traffic (photo by Michel Di Tommaso).

Stabilizing soil in a rural road. To improve its resilience, this rural road is paved with compacted soil mixed with cement and lime (photo by Michel Di Tommaso).

Such roads, however, are extremely vulnerable to intense rain and winter freezing conditions. Intense rain softens the foundations of the road and reduces the resilient modulus of pavement materials, while wintery frost conditions cause the formation of frozen lenses of water within the foundation materials, leading to potential loss of bearing capacity. Stabilizing marginal soils in rural areas has two objectives: (i) to decrease the CFP by minimizing the transport and hauling operations, and (ii) to reduce the import of materials while making the pavement structure more resilient.

D.   Adaptation Measures to Reduce Scour at Bridges

Forecasting the extent of scour on bridge piers during flood events is essential during the design stage to minimize damage. Routine inspections during dry periods after floods can help ascertain if the depth of cumulated scouring is excessive. If there is potential loss of friction between piles and surrounding soil, recommended adaptation measures include the use of riprap, geobags,[32] gabion mattresses filled with ballast materials (rock fragments, cobbles, etc.), and a collar (Figure 17).

These strategies aim to reduce the scouring effect both downstream and upstream of piers, which are among the most vulnerable parts of a bridge structure.

Example of riprap. The riprap protects the levees of a river from erosion (photo by Michel Di Tommaso).

---

32   Geobags are bags of tear-resistant material filled with sand, gravel, or other ballast materials.

Figure 17: Collar to Control Scouring

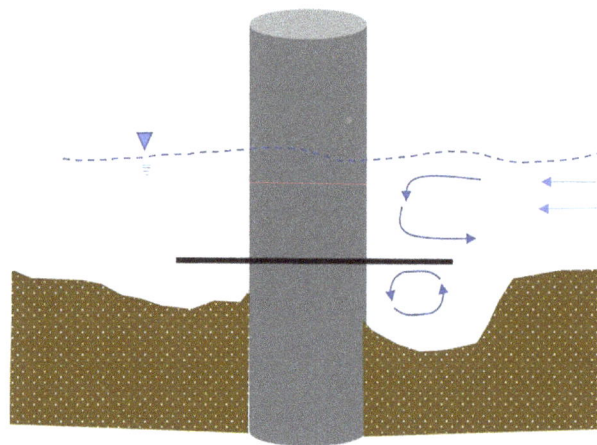

Note: The collar (black horizontal line in the sketch) breaks the downflow and reduces the horseshoe vortex intensity (blue swirls) upstream and downstream.

Source: Authors.

Example of gabion mattresses. Gabion mattresses made of steel wire boxes filled with cobbles are used for slope protection and stabilization (photo by Meaw_store, Shutterstock.com).

E.   Adaptation Measures to Reduce the Damaging Effects of Gravity Flows: Fragility Curves for Embankments

Road and rail carriageways are often the most vulnerable to gravity flows, which cause damage to transport infrastructure and disruption to traffic. They can also threaten the safety of road users when they overtop embankments with vehicles in transit. This means embankment slopes at risk of being hit by debris and/or mudflows must be strengthened and, preferably, stabilized with vegetation and/or with reinforced earth to lower the risk of erosion (Figure 18).

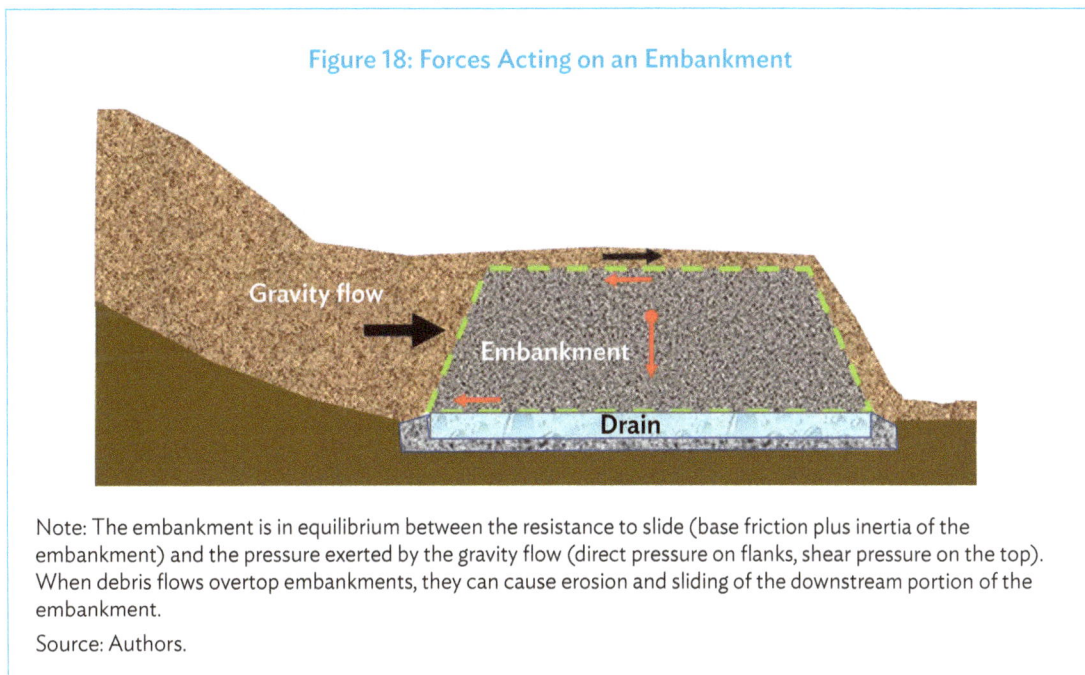

Figure 18: Forces Acting on an Embankment

Note: The embankment is in equilibrium between the resistance to slide (base friction plus inertia of the embankment) and the pressure exerted by the gravity flow (direct pressure on flanks, shear pressure on the top). When debris flows overtop embankments, they can cause erosion and sliding of the downstream portion of the embankment.

Source: Authors.

The equilibrium of the embankment is between the acting forces that cause sliding and the resisting frictional forces mobilized at the interface between the embankment and the existing ground. If the embankment is stable, it means that the resisting forces are larger. When sliding occurs, the actions have overcome the resistance. This is called the ultimate limit state (ULS) for the embankment, which means that when actions win the resistance, the structure will fail and/or slide and will potentially collapse.

This event must be made very unlikely (with a probability of occurrence of 0.1% or even less, depending on the consequences that the ULS would mean for the structure). Therefore, it is important to define the critical maximum pressures a debris flow would exert when mobilized. This requires proper geological reconnaissance and review of existing data to gather information of return periods of damaging events of this type. The acting forces when debris flows hit an embankment perpendicular to the axis can be quantified and/or estimated because they are essentially a function of the density, width, depth, and velocity of the flow as well as the internal and/or external friction at the embankment's base (Figure 19).

Figure 19: Fragility Curves for an Embankment

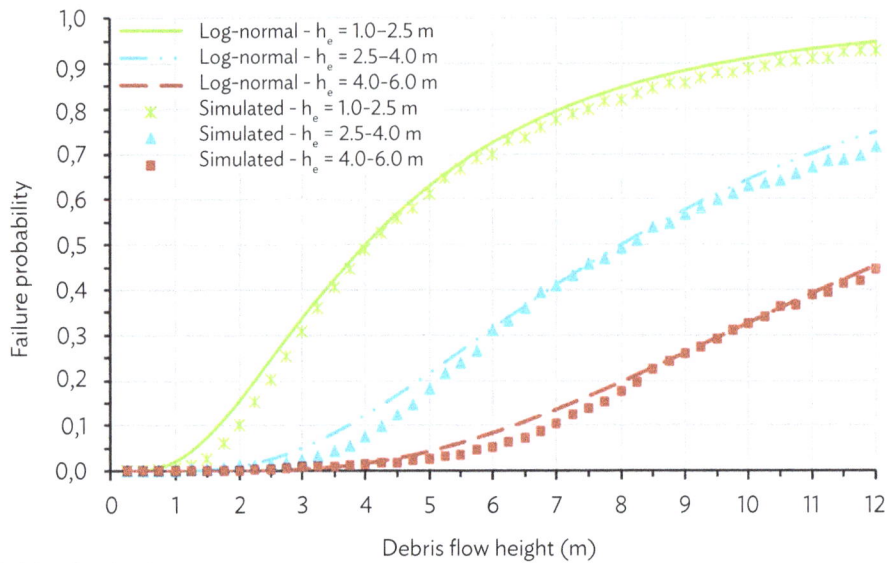

h_e = height of embankment, m = meter.

Note: For an embankment with height between 2.5 m and 4.0 m, the probability of failure is negligible until the debris flow reaches a thickness of 2.0 m. For an embankment between 4.0 m and 6.0 m, the probability of failure is negligible for a thickness of the debris flow of up to 4.0 m. Thus, increasing the height of the embankment introduces more inertia and a physical barrier to the debris flow.

Source: Authors, based on N. Nieto et al. 2021. Development of Fragility Curves for Road Embankments Exposed to Perpendicular Debris Flows. *Geomatics, Natural Hazards and Risk.* 12 (1). pp. 1560–1583.

Embankment slope. This poorly built embankment on a major highway in the Central Asian region poses risk to the safety of road users (photo by Michel Di Tommaso).

The modern approach to the estimation of the risk of reaching the ULS is based on fragility curves, which are equally valid for flood defenses. Fragility curves express the probability that a debris flow of height h_d would cause damage to an embankment of height h_e. This is done by assuming a specific rule for defining the ULS: the sum of resistances is overcome by the sum of actions. This calculation can be done iteratively many times with varying actions and assumed reduced resistances (for example, a lower than design height h_e), and then count how many times actions and resistances are such that the ULS is violated. When the calculation is done thousands of times[33] and the intensities of forces and geometrical parameters are varied to include worst-case scenarios, the number of times that the ULS is reached over the total number of trials is the probability for the condition of ULS or catastrophic failure.

Figure 18 shows that, if the area where the transport infrastructure project could be affected by debris flows and if the return period of the most damaging event is known, embankments must be built that exceed at least by 1.5–2.0 times the maximum thickness expected for the debris flow to reduce the risk of failure. When the ratio between the thickness of the debris flow and the height of the embankment is close to 1:1, the probability of failure is, in fact, more than 10%, which, in geotechnical engineering, is not an acceptable level of risk.

If geological investigations show that the transport infrastructure is at risk of damage by debris flows or mudflows, adaptation measures upstream of the road or railway must be taken to prevent the flows from reaching the carriageway with maximum energy. Some of these measures are shown in the following photos.

Energy breaking barriers for water laden sediments. These solutions allow breaking the energy of water flowing downstream during floods (photo by Akira Shimizu, Shutterstock.com).

Debris flow barriers for dry sediments. These solutions allow trapping dry or moderately wet granular materials moving downhill under gravity (photo by Michel Di Tommaso).

Debris flow barriers for water laden sediments. These solutions allow trapping granular materials (boulders, cobbles) and vegetation moving downhill under wet gravity flows(photo by Basotxerri, Shutterstock.com).

---

[33] This is accomplished by using a dedicated software allowing iterative calculations with the Montecarlo method of random sampling.

F.   Maintenance of Drainage Systems

Assuming drainage systems are properly designed for the anticipated maxima of runoff water, the only way to keep these features efficient is through adequate maintenance. Periodic and planned cleaning of drainage systems can help divert water away from vulnerable parts of the transport infrastructure (i.e., slopes, trenches, and embankments). Ditches, gutters, pipes, collection ponds, and siltation ponds, among others, must be kept free of debris and vegetal matter at all times to allow the quick discharge of rainwater.

Prevention Measures against Blockage and Damage of Drains

The most effective way to protect concrete drains from being damaged by rolling boulders during flood events is by protecting them with a cage at the inlet. This simple and cheap solution can prevent structural damage to the drain and the pavement above when the drain gets blocked and when water and debris overflow on the carriageway.

G.   Embankment Reinforcement

Embankments, which may be affected by moving debris and mudflows, can have their flanks reinforced against pressures and erosion (Figure 20). Reinforced earth systems consist of layers of suitable backfill material resting over a structural biaxial web made of steel wire or high tensile strength polymers. These are completed with gabions on the exterior. The backfill material provides inertia toward sliding, while the composite nature of the fill-web-fill system increases the shear properties of the slope as a whole. Finally, the external gabion reduces the possibility of erosion and provides further resistance against active loads. The backfill material is usually within the range of 0–100 millimeters (mm) in size.

H.   Slope Protection

Natural and artificial slopes interact with the transport infrastructure. In designing a stable slope, knowledge of the basic geotechnical properties of soils and the worst-case scenarios is needed as far as water saturation and/or fire exposure are concerned. The geometrical design of the slope (i.e., the angle of the slope and the number of berms[34]) is essential (Figure 21); however, under particular circumstances related to heavy rain, for instance, this might not be enough. Other measures should be adopted to reduce the risk of gravity-induced flows to interact with the pavement infrastructure.

Two methods for stabilizing the slopes are soil nailing (Figure 22) and using gabions (Figure 23). Hard facing includes sprayed concrete, reinforced concrete, and stone pitching.

---

[34]   A berm is a ridge or raised area built of compacted soil, gravel, stones, or crushed rocks to intercept, divert, or prevent stormwater runoff from entering a particular area. It divides an embankment into horizontal subareas to decrease the pressure on the foot of the embankment.

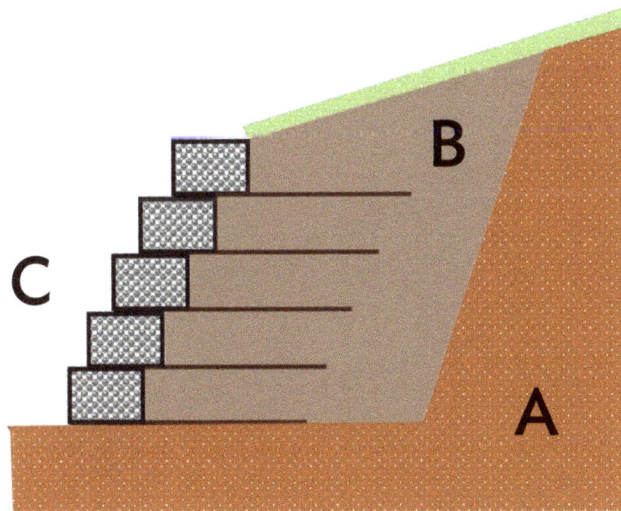

Figure 20: Sketch of a Reinforced Embankment

Note: The existing soil (A) is replaced by a suitable backfill (B) placed in layers over polymeric or a steel web (horizontal black lines), while gabion mattresses (C) are placed on the exterior to protect it from erosion and provide counterweight for possible rotational failures.

Source: Authors.

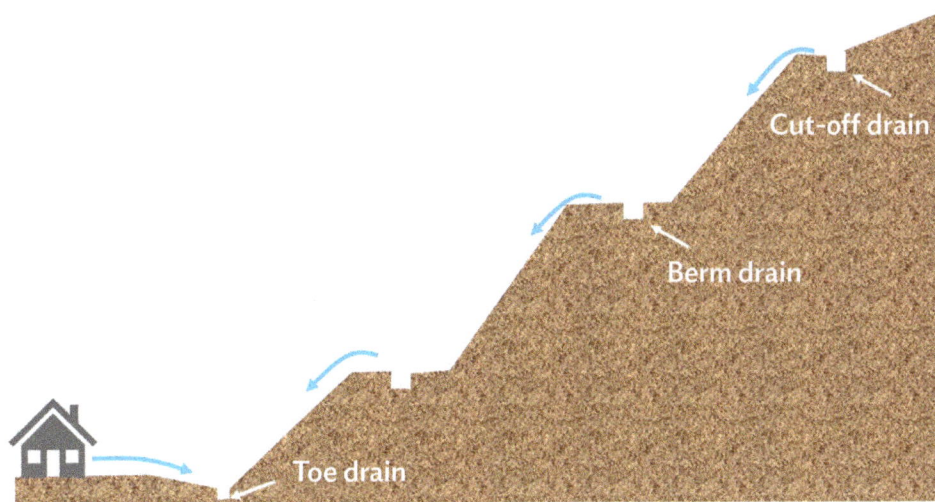

Figure 21: Sketch of a Bermed Slope

Source: Authors.

Figure 22: Sketch of a Soil-Nailed System

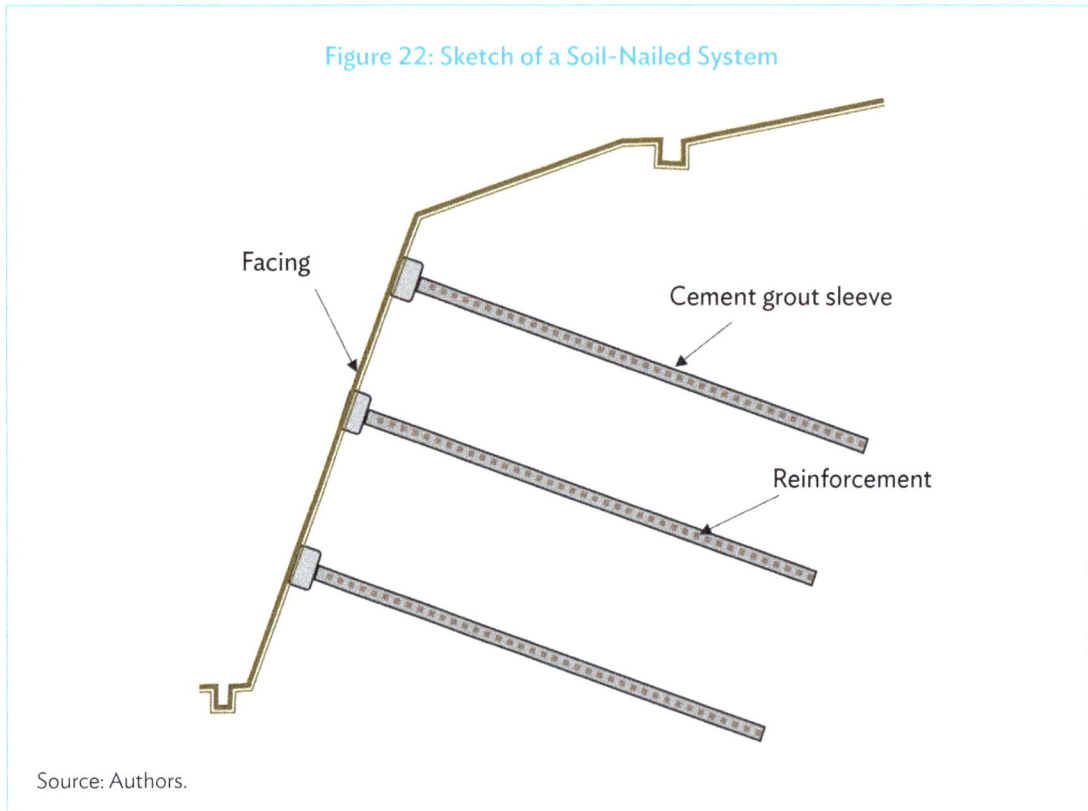

Facing

Cement grout sleeve

Reinforcement

Source: Authors.

Example of soil nailing. Soil-nailed system made with high-tensile-strength synthetic fabric is used to stabilize steep slopes and to construct retaining walls from the top down (photo by Elena Dijour, Shutterstock.com).

Gabions are gravity structures that can be placed at the foot of a slope to stabilize it and to prevent the upstream portions sliding. The basic input parameters for designing gabions are the wall's height (H), surcharge (q), slope angle, angle of friction of soil, density of soil, gabion's fill density, and soil-bearing pressure.

Figure 23: Balance of Forces for a Gabion Wall

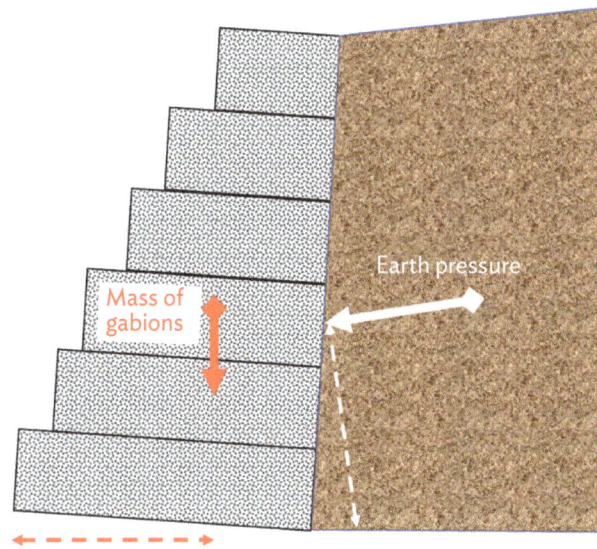

Note: The equilibrium is between the moment given by the earth pressure vector acting on the upper face of the gabions and the resisting counterclockwise moment given by the vector of inertia of the gabion wall.

Source: Authors.

Example of a gabion wall with reinforced earth. Gabion walls are typically used as retaining walls to control erosion (photo by Michel Di Tommaso).

I.   Control of Surface Runoff

In areas where the road infrastructure could be intersected by small streams, which could discharge and/or cascade high volumes of water and debris downstream onto the carriageway, the following adaptation measures can be considered:

(i)   collecting water cascading from rock faces upstream and discharging it downstream, and

(ii)   applying bioengineering solutions that can stabilize the surface of the slope and break down water runoff energy during intense rain events.

Energy breaking barrier. A spillway like solution helps trapping sediments upstream of the drainage pipe, avoiding clogging and helping maintenance operations (photo by Michel Di Tommaso).

Energy breaking bio-engineering solutions. These natural barriers reduce energy of surface run off during rain events (photo by Michel Di Tommaso).

Avalanche and rock fall protection. These solutions in mountain areas protect the road and the road users from avalanches and rock falls during winter time and during rain events (photo by Florencia Maisonnave, Shutterstock.com).

J.   Bioengineering Solutions

In its Strategy 2030, ADB pledged to promote quality infrastructure investments that are green, sustainable, resilient, and inclusive.[35] Bioengineering, as part of sustainable adaptation measures, uses plants to protect slopes and stream banks. However, it can control erosion and prevent or stabilize shallow slope movements only where the depth to failure is no more than 0.5–1.0 meter (surface phenomena only). Therefore, bioengineering solutions should be used in conjunction with conventional measures, such as gabions and reinforced earth systems (Figure 24). Bioengineering provides cost-effective methods of surface protection for slopes, embankments, and levees, which is achieved through a surface cover of vegetation that protects the surface soils against erosion.

Figure 24: Combined Techniques for Slope Stabilization

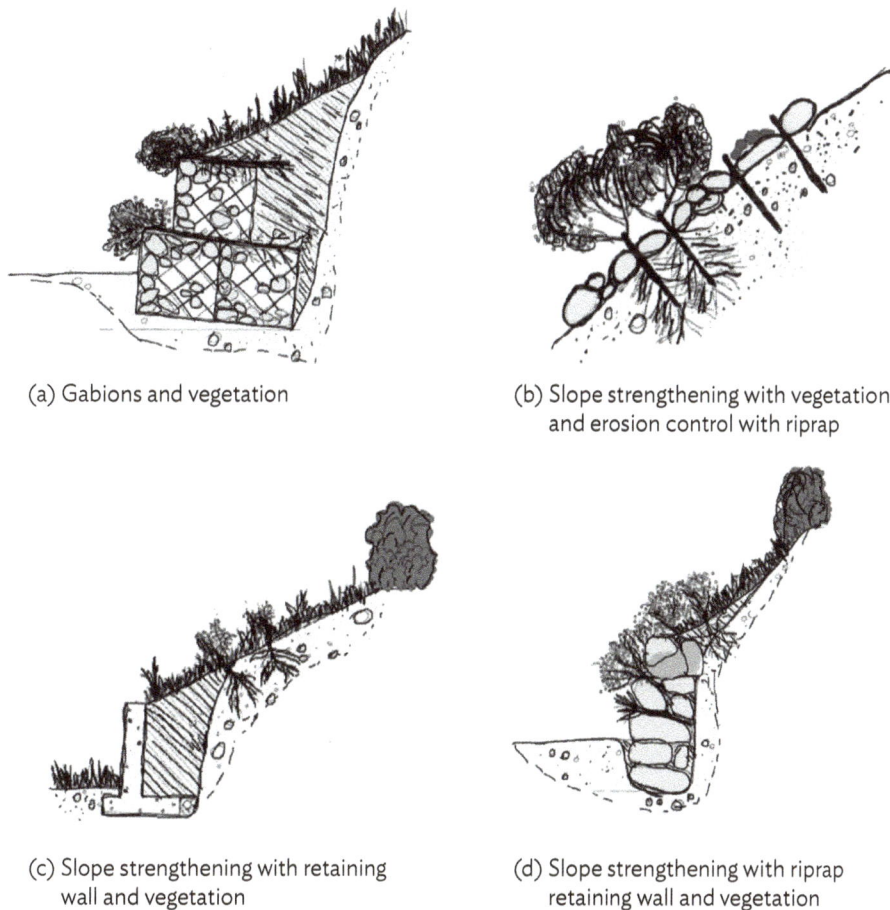

(a) Gabions and vegetation

(b) Slope strengthening with vegetation and erosion control with riprap

(c) Slope strengthening with retaining wall and vegetation

(d) Slope strengthening with riprap retaining wall and vegetation

Note: These combined techniques are also applicable to levees and riverbanks.

Source: Reworked from United States Department of Agriculture. 1992. Soil Bioengineering for Upland Slope Protection and Erosion Reduction. In *Engineering Field Handbook*. Washington, DC.

---

[35]   ADB. 2020. *Bioengineering for Green Infrastructure*. Manila.

Table 3 presents some of the bioengineering techniques appropriate to Central Asia.

Table 3: Advantages of Bioengineering Solutions

Techniques	Advantages
Grass planting: Root-shoot slip cuttings of large grasses are planted in lines across a soil slope. Slips are made by splitting the clumps to give a small section of both roots and shoots. Lines are either horizontal or diagonal, depending on slope material characteristics.	• The best and quickest way to create a surface vegetation cover on a bare slope with at least 30% soil • Effective cover on almost all soil slopes up to 2 vertical: 1 horizontal • Robust protection and shallow reinforcement of the surface soil • Grass may also be seeded over bare surfaces, but this frequently loses all engineering functions other than armoring.
Direct seeding: The seeds of shrubs and small trees are inserted into crevices in slopes made of moderately weathered rock.	• The best way to establish vegetation on rocky slopes
Brush layers and fascines: Wood cuttings from shrubs or small trees are laid in different ways on shallow trenches across slopes formed in unconsolidated debris. They can be installed on slopes of up to about 1 vertical: 1.25 horizontal. Palisades can be on steeper sites, up to about 1.75 vertical: 1 horizontal.	• Instant physical barrier that interrupts runoff and is stronger than grass • Strong erosion protection and soil reinforcement as the plants take root and grow • Often successful on stony debris • Most shrubs will tolerate some shade, so this method can often be used under tree canopies where grasses will not grow.
Truncheon cuttings: Big wood cuttings from trees are inserted upright at intervals on slopes or stream banks formed in deep or poorly stabilized and unconsolidated debris.	• Relatively strong plant material on slopes that are still unstable • Ability to withstand damage from moving debris
Live check dams: Small check dams with structural elements made from wood cuttings from trees or bamboo are placed at intervals in erosion gullies.	• Low-cost, flexible structures to reduce erosion where water flow is concentrated but ephemeral • Relatively limited disturbance to the slope, particularly on weak, unconsolidated materials
Tree planting: Potted seedlings from a forest nursery are planted at intervals across a soil slope.	• Restoration of a forest mix of trees in the long term • Trees take at least 5 years to contribute significantly to slope strengthening or to establish a complete cover, and initially may be vulnerable to grazing.

Source: Asian Development Bank. 2020. *Bioengineering for Green Infrastructure*. Manila.

K.   Adaptation Measures to Reduce the Temperature Effects on Pavements

Introduction

Both asphalt and concrete are thermally sensitive to heat and cold. For concrete, the sensitivity is seen through the expansion and contraction of the pavement structure experienced at the change of season or with day and night cycles in continental climates. For asphalt, the effects are related more on the loss of stiffness of the material itself at high temperatures.

Unlike concrete, when asphalt heats up, it no longer behaves like an elastic body where the strain is proportional to the stress. When asphalt is hot, it behaves like a viscous-plastic medium, where the strain grows faster than the applied stress, and some degree of plastic (or unrecoverable) strain remains stored during load applications. The result of this accumulation of plastic strain is, in fact, the rutting phenomenon observed in the summer on many asphalt roads subjected to heavy traffic. In simple terms, if rutting is viewed as a phenomenon of plastic failure and shear along preferential planes of weaknesses within an asphalt layer, the governing equation for the equilibrium can still be expressed as follows:

$$\tau = \sigma tan\varphi + c$$

where

τ = shear strength of the asphalt concrete

σ = vertical stress

φ = angle of friction of asphalt

c = elastic threshold to be overcome for failure to occur (i.e., internal cohesion provided by the binder)

In cold temperatures, the above equation holds entirely. In hot temperatures, the cohesion term drops to zero and the shear strength of the material decreases accordingly. Hence, it takes smaller shear stresses to cause failure (or to cause high shear strains) of asphalt when it is hot. When hot asphalt fails, it produces rutting and lateral flow, making driving dangerous because of aquaplaning.

Adaptation Measures against Extreme Heat for Asphalt Roads

To make asphalt more resistant to softening in the summer, the bitumen modification process is applied. Bitumen modification is a method for engineering the properties of natural bitumen to make it retain its stiffness during summer and remain ductile during winter and flexibly expand and contract during thermal cycles without cracking.

The modification of bitumen is made through polymeric fibers and with proprietary methods developed by fiber manufacturers. The modification can be made at the source, and bitumen is sold modified-ready and is incorporated in asphalt concrete with aggregates. Less frequently, it can be modified at the batching plant, where polymeric fibers are mixed with virgin bitumen and aggregates during the production of asphalt concrete.

This modification enhances the shear strength of the material when the asphalt is hot in service. By losing its elastic behavior when heated up, asphalt behaves like an unbound cohesive soil that can only resist deformation imposed by traffic loads through the shear strength. This is because the binder, when heated, loses the cohesive forces holding the grains together, and grain-to-grain friction prevails. Adding fibers increases the shear strength of asphalt concrete by restoring the internal cohesion term that would otherwise be lost when virgin bitumen is heated up. Fibers will work against the tensile and shear forces forming in the hot asphalt when loaded by traffic. The shear and tensile resistances provided by the fibers will effectively reduce rutting.

Table 4: Performance Grade Classification of Bitumen

	High temperature (°C)				
Low temperature (°C)	52	58	62	70	76
-16	52-16	58-16	54-16	70-16	76-16
-22	52-22	58-22	62-22	70-22	76-22
-28	52-28	58-28	62-28	70-28	76-28
-34	52-34	58-34	62-34	70-34	76-34
-40	52-40	58-40	62-40	70-40	76-70

Crude oil   High-quality crude oil   Modifier required

Note: Performance Grade (PG) 66, PG72, are examples of bitumen grades suitable for the climate of Central Asia and for heavy traffic.
Source: Authors.

When bitumen is modified by a suitable additive to increase shear strength, the bitumen is called performance grade. Table 4 shows the different performance grades for bitumen in relation to maximum and minimum temperatures. In continental climate conditions, where very high and very low pavement temperatures are expected, a modification is always required. Hence, for continental climates such as in the vast majority of Central Asia, modified bitumen should be preferred over virgin bitumen to enhance stiffness in the summer, without showing an excessively stiff and rigid response in the winter that would make it prone to thermal cracking.

Adaptation Measures against Extreme Heat for Concrete Roads

For concrete roads, the maximum contraction and expansion need to be predicted in relation to the moment concrete is laid. Expansion joints are used to cope with the control of blow-up pressure caused by two adjoining sections of a concrete road. Furthermore, joints should not open excessively in winter, because the load transfer efficiency between two adjacent portions of concrete is reduced the further apart the two sections are.[36]

The general concept behind the use of expansion joints is to always allow a gap between the two sections of rigid pavement, even when in maximum summer expansion. This requires making engineering considerations on the coefficient of thermal expansion of the concrete used, the friction with the base, the maximum possible summer and winter temperatures, and the duration of the exposure of the concrete surface to them.

---

[36]   ADB. 2021. *A Practical Guide to Concrete Pavement Technology for Developing Countries*. Manila.

The larger the temperature difference between the moment of casting and the maximum temperature in service, the smaller the expansion joints' spacing should be (see the Worked Example on p. 34). When contraction joints (used at 4-meter to 6-meter intervals to prevent the shrinkage of concrete pavement) are blocked because of debris flushed on the pavement by wind and rain, expansion joints carry even larger compressive stresses than during the summertime pavement expansion. Hence, to adapt to climate change in terms of stress relief joints in concrete pavement, the design of expansion joints must be in a way that more expansion is accommodated in the summer. This means that more expansion joints per unit length of road must be installed to keep the opening of the joint to a maximum of 30 mm. The typical section of an expansion joint is shown in Figure 25.

Figure 25: Detail of an Expansion Joint

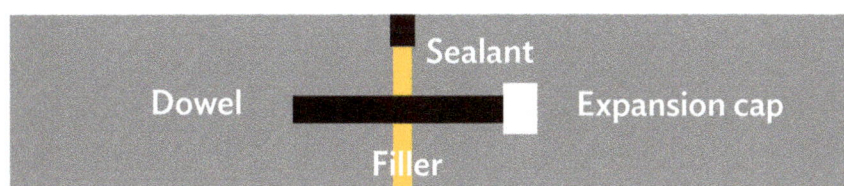

Note: Dowels (black) are used to transfer loads between the two sections of concrete, which are separated by a compressible filler (orange). The dowel is free to follow the expansion and contraction of the sections via the expansion cap.
Source: Authors.

Adaptation Measures against Snow Piling

Snow that accumulates on naked and/or steep slopes will eventually mobilize and flow downstream with potentially damaging consequences for road infrastructure. To reduce the risk of high volumes of snow and debris impacting structures, slopes are stabilized with barriers made from steel or wood to increase the friction between the base of the snowcap and the ground and to contain and/or reduce the downflow of wet snow masses.

Avalanche hazard adaptation. A steep and naked slope is fenced with proprietary bioengineering solutions for controlling snow avalanches (photo by Mario Krpan, Shutterstock.com).

L.   Adaptation Measures against Rockfalls

Rockfalls occur when heavily fractured rock masses have a favorable orientation to slide. When water is present and/or when the binding force of ice filling the cracks is lost by warming (such as permafrost melt), the slopes made of fractured rocks become unstable. These rocks slide, topple, or fall, causing potential damage to both the road infrastructure and its users (Appendix 4).

The best options avoid blocks rolling downstream by using attenuators, canopies, or drapes. These features act as shock absorbers.

Protecting against rockfall. A rockfall attenuator is built to protect a road (top photo) and a railway line (bottom photo) (photo by Michel Di Tommaso).

Focus on Uzbekistan

5

A. General Political Framework for Action

In Uzbekistan, the Ministry of Transport primarily administers the development and implementation of a unified state policy focused on developing roads, railways, air transport, river transport, metro, and road facilities. Accordingly, the ministry adopts regulatory acts; issues licenses, permits, and certification; and enforces an effective technical and tariff policy. The State Committee for Roads, under the Ministry of Transport, is responsible for the following tasks:

(i) implementing a unified technical motorway policy;
(ii) developing and implementing state programs for the development of motorways;
(iii) determining the prospects for the development and improvement of the road network;
(iv) forming international transit corridors for motorways;
(v) providing a comprehensive solution to the financing, design, construction, repair, and operation of motorways, taking into account the interests of road users as well as the conditions of modern traffic flows, and organizing effective customer service activities;
(vi) controlling the quality of construction, reconstruction, repair, and maintenance of motorways;
(vii) coordinating works to ensure the safety of the existing network of interfarm rural motorways, city streets, urban settlements, and villages, while maintaining them at a high transport and operational level;
(viii) organizing research work and implementing innovative technologies and modern standards in the design, construction, reconstruction, repair, and maintenance of motorways; and
(ix) organizing training, retraining, and advanced training of personnel in the field of motorways, including conducting training courses and seminars abroad.

Based on the above functions, it is appropriate that the Committee for Roads acts as the lead agency in climate change adaptation (CCA) measures. This could involve measures to achieve the following:

(i) incorporating CCA measures in road infrastructure design standards;
(ii) producing directives to incorporate CCA measures at the project preparation stage and mainstream them in the planning, design, and implementation stages;
(iii) establishing a CCA office under the Committee for Roads to mainstream CCA in design and construction;
(iv) recruiting qualified technical personnel knowledgeable with CCA;
(v) providing sufficient technical support to the CCA office to adequately address the corresponding project requirements—technology, software, among others;
(vi) undertaking countrywide risk modeling in the transport sector to identify the hot spots for building resilience, delving into all possible types of hazards (as practiced already in a number of countries);

(vii) establishing a robust system to collect data on the impact of climate extremes on roads to help improve future planning, in coordination with other allied agencies;

(viii) operationalizing ex ante contractual arrangements for quick inspection and repair after the occurrence of an event; and

(ix) building capacity of the broader construction sector to incorporate CCA issues and concerns in road design, implementation, and maintenance.

In addition to the Committee for Roads, a number of state agencies can contribute to speeding up the adoption of CCA measures in road infrastructure projects. These agencies are as follows:

Ministry of Emergency Situations. As the government agency overseeing emergency services in Uzbekistan, it is responsible for providing emergency assistance to people, protecting them during disasters, overseeing emergency measures, and coordinating with other ministries and departments. As such, this ministry would be in a good position to stipulate the requirements for the safety of infrastructures in the face of potential climate change risks such as flooding, slope collapse, avalanche, and wildfires. Hence, more stringent measures to avert emergency situations can be part of the requirements in line with this ministry's purview. Vulnerability assessments for the entire road network can be initiated by this agency in coordination with the Committee for Roads. Such assessments can provide a general overview of the potential risks to infrastructure linked to climate change. The ministry can also enable early warning systems (EWSs) supported by contingency plans for emergency response, rerouting, and recovery, which can have improved connectivity with the citizenry in a crowd-sourced information system. This EWS can be further enhanced with a cyber-based decision support system, which can be interactive when accepting and processing information that can enhance the response capabilities of the ministry. In this way, the public becomes a partner in providing timely information for the EWS.

State Committee for Ecology and Environmental Protection. Responsible for ecology, environmental protection, and the use and improvement of natural resources, the state committee reports to the Cabinet of Ministers of the Republic of Uzbekistan. The state committee enforces the environmental policy that aims to create favorable conditions for environmental safety and protection of the country, improve the environmental situation, prevent the harmful impacts of waste on the environment, and improve the quality and standard of living. It is also the primary authority for processing environmental impact assessments through the unified system of the State Environmental Expertise. In line with environmental safety, CCA should form part of the fundamental requirements within the entire process—from project conceptualization, planning, and design, to implementation. For areas with certain elevated geohazards, a more thorough assessment may be necessary. This necessitates a more comprehensive engineering geological and geohazard assessment as an additional requirement for environmental permit. The measures in these assessment reports should be specific to addressing the geohazards identified. This additional requirement can serve as a good guidance for the projects being implemented. Likewise, this agency can spearhead the promotion and upscaling of eco-based approaches, including bioengineering, as supplementary requirements in infrastructure projects. Thus, it can prescribe project proponents to study and assess the viability of combining nature-based solutions with structural measures.

State Committee on Land Resources, Geodesy, Cartography and State Cadaster. This body is responsible for implementing a single state policy in the field of land resources, geodesy, cartography, and state cadaster. In line with this primary function, the agency should undertake hazard mapping for the entire country and identify regions and areas that will be subject to more intense assessment with respect to CCA requirements.

Research Institute Center of Hydrometeorological Services (Uzhydromet). Under the Ministry of Emergency Situations, this state governing body is authorized to deal with hydrometeorology in Uzbekistan. The objectives of Uzhydromet are to develop and improve the state system for hydrometeorological observations, conduct scientific research activities, improve short-term and long-term weather forecasts, and study water availability and climate change in general. With its dedicated function, Uzhydromet will be at the forefront of providing information for predicting and modelling extreme weather events to help improve infrastructure design and capacity.

B.   Climate Change in Uzbekistan

Located in the heart of Central Asia, Uzbekistan's terrain is mostly flat-to-rolling sandy desert with dunes. Its eastern and northeastern regions consist of mountains and foothills. The climate is typically arid continental, while the northern region is temperate and the southern region subtropical. The majority of the country's area has a moderate climate characterized by seasonal and day-to-night fluctuations in air temperatures. In summer, the average July temperature on the plains' territory reaches the upper 20s°C and the upper 40s°C. In winter, the average January temperature falls below 0°C. Most of the precipitation primarily occurs during the winter–spring period. Variations in annual precipitation depend on the area. It can average 80–200 mm in the plains, 300–400 mm in the foothills area, and 600–800 mm in the eastern and southeastern slopes of the mountain ridges.

Uzbekistan, like most countries in Central Asia, has been experiencing pronounced warming since 1950. Based on the studies of the National Communications of Uzbekistan, supported by the United Nations Framework Convention on Climate Change, climate change is expected to cause the following:

- The boundary between the arid tropical and temperate climatic zones will shift north by 150–200 km, and the altitudinal climatic zones will shift upward by 150–200 m.
- Duration of the frost-free period will increase by up to 15 days.
- Temperatures will increase by 5%–10%.
- Climate change will significantly affect the volume of glaciers feeding most of Uzbekistan's rivers.
- Increased intensity of rain events and soil depletion and erosion will increase the sediment load of rivers, resulting in further sedimentation of irrigation channels and water reservoirs.
- Mudflows will likely increase in number and intensity.
- Warming in the mountainous zone will lead to higher risk of lake outbursts.

These risks could have wider impacts and expose inherent vulnerabilities. For example, disruption to transport has a direct impact on the economy and can hinder access to employment, education, and health.

Regional Vulnerability Related to Climate Issues: The Cases of Aral Sea and Sardoba Dam

The hydrologic status of the Aral Sea has been subject to numerous studies. Prior to the 1960s, the Aral Sea, with an area of 68,000 km², was listed among the four largest lakes in the world. With the diversion of water for irrigation during the former Soviet Union, the Aral Sea started shrinking and, by 1997, it had shrunk to 10% of its original size. The decrease in size also split the waters into smaller basins—the North Aral Sea, the eastern and western basins of the once far larger South Aral Sea, and one smaller intermediate lake. By 2009, the southeastern lake had disappeared, and the southwestern lake had retreated to a thin strip at the western edge of the former southern sea. In August 2014, images from the National Aeronautics and Space Administration (NASA) showed that the eastern basin of the Aral Sea had completely dried up, giving rise to a new desert, the Aralkum Desert. The region's once prosperous fishing industry has been essentially destroyed, bringing unemployment and economic hardship. Although huge areas of desert have been converted into farmland within Uzbekistan, inefficient irrigation and drainage systems are causing large losses of water. With the increase in summer temperatures over the years due to climate change and with decreasing water sources, the subsoil is increasingly drying, giving rise to elevated salinization, a decrease in vegetation, and the spread of desertification. The decrease in vegetation is putting a strain on biodiversity, and the decline in agricultural productivity brings economic difficulties to the regions within Uzbekistan and neighboring Kazakhstan.

With ADB's assistance, through the Climate Adaptive Water Resources Management in the Aral Sea Basin Program, irrigation and drainage in selected subprojects within the Amu Darya River and reaches of the Zarafshan River Basin in Uzbekistan will be modernized. The project aims to deliver climate-adaptive solutions to water resources management toward improved food and water security in the Aral Sea Basin in Uzbekistan. The project outputs are as follows:

(i)   establishing a climate-resilient modernized irrigation and drainage subproject,
(ii)   providing enhanced and reliable on-farm water management, and
(iii) strengthening policy and institutions for sustainable water resources management.

Adaptation measures in irrigation and drainage infrastructure will follow the build back better approach wherein civil works will be conceptualized, designed, and constructed with due consideration to extreme weather events to increase the resilience of infrastructure.

In the morning of 1 May 2020, after a week of heavy rains in Uzbekistan's Sirdaryo and Jizzakh regions, an embankment of the earth-filled dam collapsed, sending rampaging waters to cotton fields and villages, and the waters spilled over to areas in neighboring Kazakhstan. This inundated more than 35,000 hectares of land within Uzbekistan and Kazakhstan. As reported, six people lost their lives and at least 111,000 people were evacuated from within the Syr Darya River Basin. Initial estimates showed that the cost of recovery exceeded the cost of constructing the reservoir. Based on ADB's initial findings, the disaster's effect was primarily in transport, agriculture, and housing.

The damage caused by the event was appalling, since the infrastructure was conceived to improve the lives of the people in the projected service area. With the worsening climatic events, such as heavy rainfall, this event is not entirely unexpected. Such incident proves that it is imperative for design engineers to consider worst-case scenarios that can serve as a basis for adopting CCA measures. Reconstruction of the Sardoba Dam's primary infrastructure should therefore consider the build back better approach.

C.   Projected Climate-Related Issues in Uzbekistan

As the impacts of climate change are increasingly felt, Uzbekistan, like other countries in Central Asia, is expected to experience intensified mudflows and debris flows caused by extreme rain following long dry spells. Other effects are likely to include an increase in rock mass instability because of glacial and permafrost melt; increased wet snow piling during the winter and greater risk of disruptive avalanches; a rise in the number of wildfires causing slopes to remain without vegetative cover and, thereby, prone to sliding; and an increase in the maximum temperature by several degrees, which could soften asphalt and cause the damaging expansion of concrete pavement.

It is therefore critical that the government and the local engineering community are aware of and alert to the risks of not implementing suitable and resilient adaptation measures today to cope with the changing climate of tomorrow.

The Financial Impact of Climate Change in Uzbekistan

There are no cheap solutions to cope with the impacts of climate change because increasing the resiliency of the built infrastructure and building more resilient infrastructure require designs to be more conservative and, thus, more costly. However, the cost must be offset against the potential damage and loss of life that could be caused by increasingly intense events.

Adaptation and mitigation will mean increased costs for transport infrastructure maintenance and construction in the years to come. Regardless of the cost of building a resilient infrastructure, periodic maintenance is the only way to extend the service life of structures. The present-day investments to improve climate change resilience, expressed as a percentage of gross domestic product for countries in Central Asia, are shown in Figure 26.

The European Commission states that, "If no new policy measures are adopted to combat global warming, the cost of climate change in Europe could reach almost 4.0% of the gross domestic product of the European Union by the end of the century."[37]

The worsening impact of climate-related stressors on transport infrastructure in the coming decades will force governments globally to allocate bigger budgets to CCA, with poorer countries forced to divert funds from other essential budget categories (health, education, research, etc.) and seek for loans from donors. All of the adaptation measures described in this manual require investment, and only limited cases resort to homemade bargain solutions such as bioengineering. However, determining the vulnerability of the infrastructure is based on the concept of risk. An event may be

---

[37]   A. Giordani. 2014. Cost of Climate Change in Europe Could Reach 4% of GDP. Horizon. European Commission. 22 January.

Figure 26: Percentage of Gross Domestic Product Invested for Adaptation by Countries in Central Asia

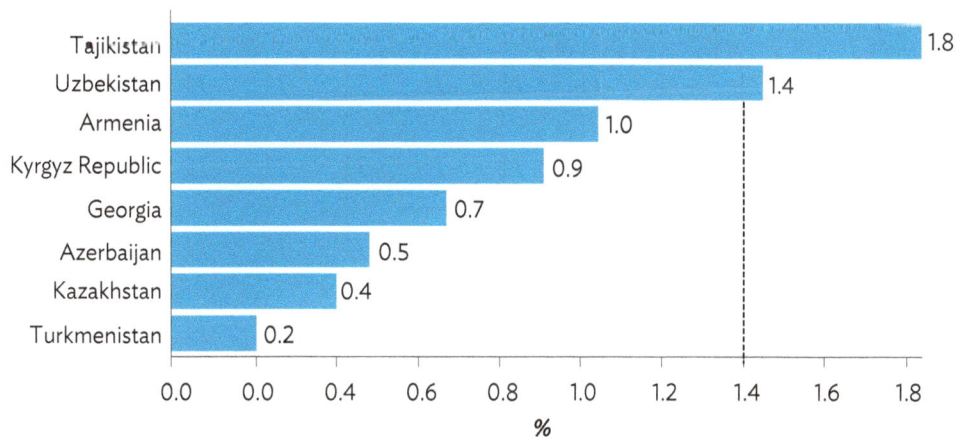

Country	%
Tajikistan	1.8
Uzbekistan	1.4
Armenia	1.0
Kyrgyz Republic	0.9
Georgia	0.7
Azerbaijan	0.5
Kazakhstan	0.4
Turkmenistan	0.2

Source: The EU Research and Innovation Magazine.

unlikely but extremely disruptive; hence, the risk factor is high. On the other hand, an event may be common and less intensive, shifting the risk factor to low. Thus, only when the risk factor is low can "lightweight" measures be adopted, which can include locally available bioengineering solutions. As the level of risk increases, more structurally sound resistances must be built, integrating bioengineering with other designs such as gabions and earth reinforcement.

High-risk factors demand technical solutions that bear a higher cost, which cannot be waived, especially when the potential consequences are damaged infrastructure and loss of lives.

D. Maintenance and Service Life

The service life of a structure is the time elapsed between construction and the structure being no longer fit for use. A road becomes unfit when, for example, at least 10% of the surface is damaged. For reinforced concrete, unfit is when corrosion of rebars begins to appear to the naked eye. To anticipate how quickly the serviceability limit state (SLS) is going to be reached, periodic inspections are needed.

If nothing is done after the SLS is reached, structures will progressively deteriorate more and more until failure. In this case, structures have reached the ultimate limit state (ULS). Roads and reinforced concrete structures are designed to meet certain service lives. Roads are generally designed for 20 years with asphalt and for 20–30 years with concrete, while reinforced concrete structures are designed for a minimum of 25–100 years or more, depending on the function. Building structures with extended service life reduces the maintenance required to keep the structure fit. Reducing the number of interventions on a pavement by designing it to be more resilient also reduces its carbon footprint (CFP).

A drawback of resilient designs are higher initial costs because resilient structures require larger and thicker slabs, beams, columns, and piers; soil improvement; slope protection; higher concrete grades; use of modified bitumen for producing asphalt; and the introduction of expansion joints in concrete pavements. Finally, when a resilient design includes calculations and comparisons between the embodied carbon in the material and the processes selected and, where possible, includes afforestation schemes, it becomes a sustainable design.

E.   Adaptation Measures for Uzbekistan

Introduction

Planning adequate adaptation measures relies on geomorphological, geological, and hydrological surveying. Periodic inspection plans are also needed to ensure that early signs of reaching the SLS are duly detected and, when required, a repair and/or refurbishment plan is implemented to reduce the risk of premature failure of the structures. Finally, EWSs are needed to gather and integrate early information on possible catastrophic events so both infrastructure and its users can be protected.

Geological Reconnaissance

The physical environment of Uzbekistan is variate, ranging from a flat topography that comprises almost 80% of the country's territory to mountain peaks in the east reaching about 4,500 meters above sea level. The southeastern portion of Uzbekistan is characterized by the foothills of the Tian Shan mountains that rise higher in neighboring Kyrgyz Republic and Tajikistan and form a natural border between Central Asia and the People's Republic of China.

Its transport infrastructure is most vulnerable in the piedmont regions bordering the mountain chains in the southeast and northeast of the country. Piedmont regions and foothills are the origin of the vast majority of disruptive events, such as floods and gravity flows, because of steep and unstable slopes, gorges, canyons, and other morphologic features that can trigger or even magnify the intensity of these events. Hazard and geomorphological maps issued by relevant authorities provide useful information on the dynamics of landscapes.

Early Warning Systems

EWSs are a set of tools, mainly instrumentation, which can be used in catchment and other vulnerable areas to monitor the dynamics of the landscape and gather information on possible sliding and other gravity flows.

A typical EWS includes

(i)   meteorological stations that monitor temperature, humidity, radiation, snow height, and rainfall;

(ii)   channel monitoring systems, such as digital cameras, located in key positions along the paths of known areas of vulnerability;

(iii) radar distance sensors that record land deformations and movements;

(iv) rotational laser scanners;

(v)   photogrammetry to show morphological changes over time, and

(vi) drone surveys.

Periodic Inspection

Periodic inspection is required to make sure that any part of the transport infrastructure is fit for use and has not shown signs of reaching the SLS. Fitness for use depends on the parameter to be used in defining whether a structure is fit or not. For an asphalt or concrete road, for instance, the pavement is considered unfit when the percentage of distress (usually rutting, cracking, raveling, etc.) exceeds some defined value—usually more than 10% of the surface.

Based on the reports of trained inspectors, asset managers should be able to determine the current state of a structure and define a suitable maintenance plan according to the amount of deterioration recorded in comparison with prior inspections. Reports may include visual condition assessment and/or nondestructive tests. In the visual condition assessment, the structures are evaluated with standard ratings and scores, allowing the user to build progressive categories of damage. These old-fashioned methods are time-consuming and are sometimes dangerous because inspections usually happen with open traffic on roads. Nondestructive tests, on the other hand, are special types of tests that can be done on the structures to quickly determine the key indicators for fitness.

Damages on a concrete pavement. The damaged joint sealant (left photo) and polished surface (right photo) of a concrete pavement pose a threat to road safety (photos by Michel Di Tommaso).

To assess whether a road has sufficient bearing capacity, for instance, a very common type of nondestructive testing uses a heavy-weight deflectometer. This equipment can load the pavement with a 50-kilonewton[38] rammer falling over a predefined area in the pavement, while displacement sensors, located at some fixed distance from the impact zone, will record how the pavement vibrates at distances from the loaded area. Plotting the vertical displacements versus the distance from the loaded area shows how the displacement is maximumly close to the loaded area, and decreases with distance from the impact (Figure 27).

Concrete, asphalt, and granular soils respond as linearly elastic materials to dynamic loads. All these materials, except for asphalt in hot weather, will therefore vibrate elastically under the impacting load. The way displacements will attenuate with the distance from the source will essentially depend on the thickness of the layers and on the resilient modulus of each layer.

The shape of the curve joining all the values of deflection recorded during a heavy-weight deflectometer test is called a deflection basin. Stiffer and more rigid layers within the pavement cause the deflection basin to be smoother and less pronounced, while weaker and more deformable layers (for instance, because of water saturation or fatigue stresses exerted by traffic over the years) have more pronounced deflection basins. From the shape of the basins and from the algorithms used to back-calculate the mechanical properties of the layers that best fit the observed displacement profiles, it is possible to estimate with good precision the in situ resilient modulus of the pavement materials and to assess whether the bearing capacity of the road is acceptable or not for the level of traffic in service.

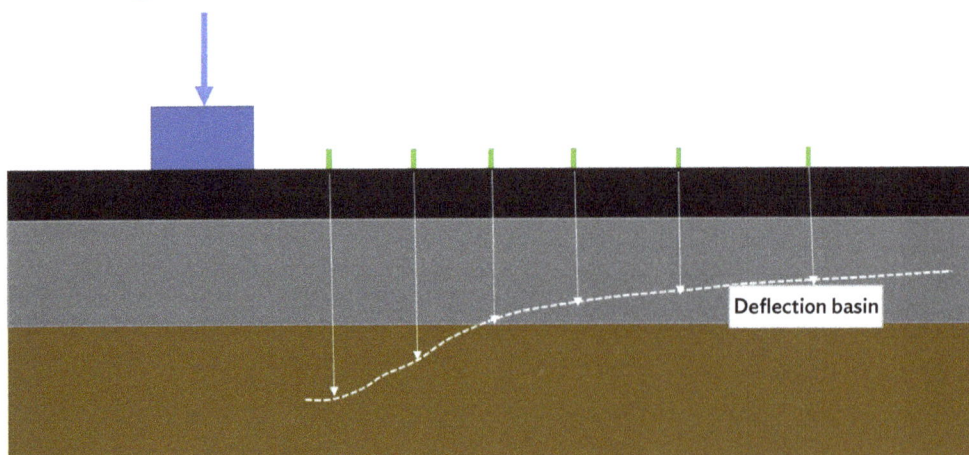

Figure 27: Deflection Basin in a Heavy-Weight Deflectometer Test

Deflection basin

Note: Sensors (thin arrows) record the vertical displacement induced by the load in various points at distance from the load (thick blue arrow). The shape of the basin (white dotted line) provides information on the stiffness of materials within the pavement's structure.

Source: Authors.

---

38   The weight varies and it depends on the type of pavement and the parameters sought (1 kilonewton = 1,000 newtons = 100 kg).

Another advanced tool for assessing the road condition is the laser scanner. These are mounted on instrumented vehicles and provide imaging of the surface of the pavement, highlighting cracks, potholes, rutting, deformations, and gaps at joints, among other factors. By periodically inspecting the surfaces using such techniques, it is possible to visualize which areas of the pavement have undergone rapid deterioration.

The advantage of laser scanners is that they can acquire all the information while the equipment travels at normal speed on the carriageway and do not interrupt traffic.

F. New Digital Capacities for Due Diligence and Pre-Assessment of a Road Project in Uzbekistan

Introduction

Infrastructure projects are complex and include multiple variables that will have a major impact on the sustainability and the resilience of the built infrastructure. Each road design must be optimized to cope with local conditions. For example, optimization for pavements requires defining thickness and mechanical properties of the layers making up the pavement structure. The modifications must also consider their impact on the CFP as part of a mitigation strategy (mandatory).

New digital tools allow engineers to perform due diligence and pre-assessment on projects based on local data gathering and analysis. Comparative analysis on multiple scenarios with multidimensional parameters—thanks to digital simulations—offers the ability to estimate the response and suitability of certain pavement structures around a few key indicators, such as

- (i)   pavement solutions and mechanical performance over the life cycle;
- (ii)   circular economy options;
- (iii) carbon emission and environmental impact;
- (iv)   overconsumption of valuable resources (e.g., water estimates for the road construction); and
- (v)   climate change impacts and main risks to the project, such as heat or floods.

These novel tools support asset managers in their decision-making and make comparisons visual and simple, even for a nonspecialist.

Case Study for Uzbekistan

Supported by ADB's digital sandbox program, the Government of Uzbekistan's Committee for Roads is using a new digital tool to carry out due diligence on a project that involves using concrete to repair a 25-kilometer section of the A380 highway.[39]

---

[39]   This is in reference to the Guzar–Bukhara–Nukus–Beyneu 25-kilometer section.

It allowed multiple-scenario digital analysis to help users make informed and data-driven decisions at an early stage, while maximizing the structural performance of the pavement (Figure 28).

Figure 28: Four Scenarios for a 25-Kilometer Section of the A380 Highway

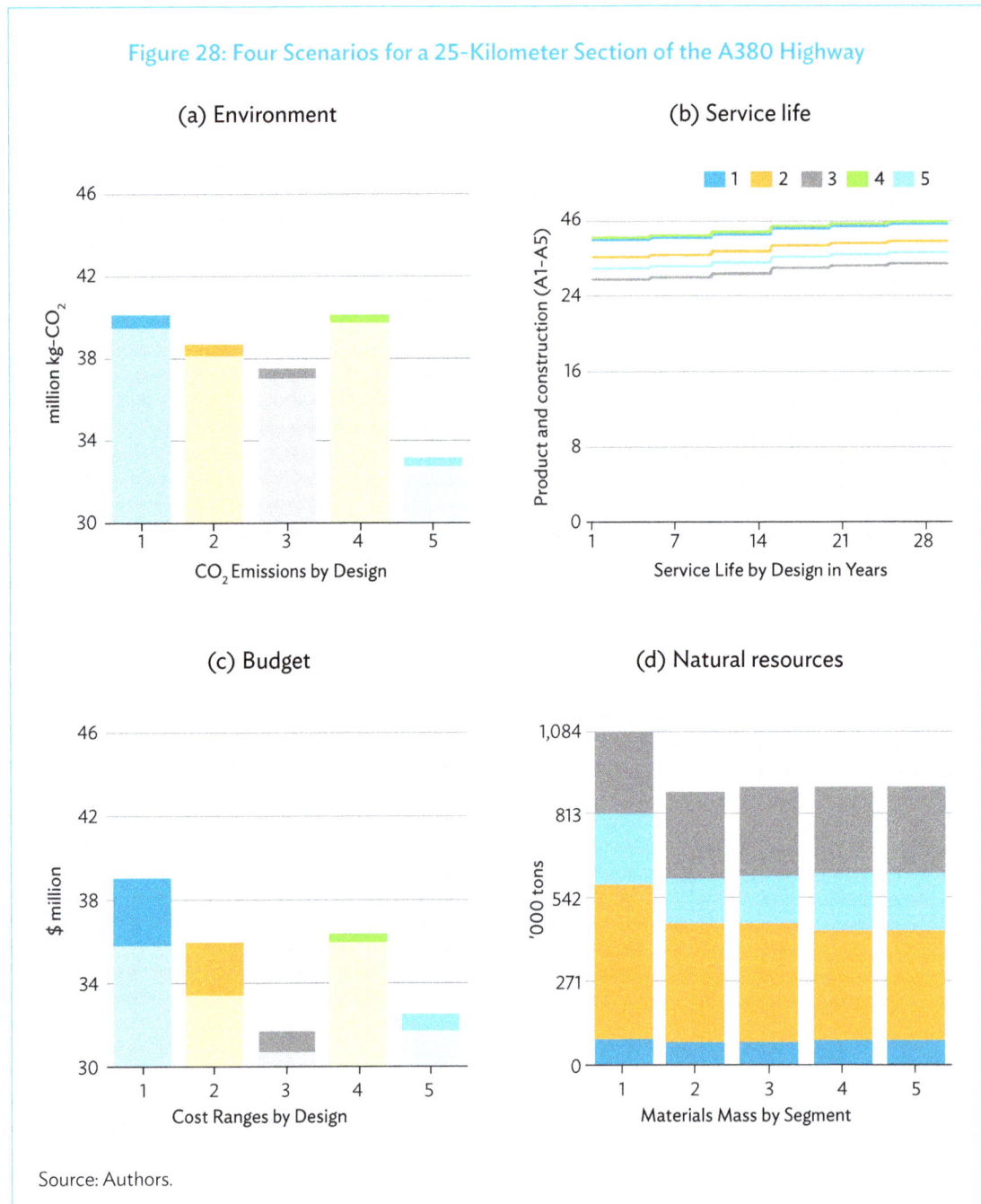

(a) Environment

(b) Service life

(c) Budget

(d) Natural resources

Source: Authors.

After compiling and processing all the information on materials, construction, and local conditions data around the project, the results provided by the digital tool could be summarized as follows:

- The pavement structure could benefit from a recycled layer, which would contribute to carbon saving and low-carbon material industrial production of 18%.
- Five million liters of water could be saved in an area with known water shortages.
- Climate change analysis covering 40 years showed how the project will be impacted in terms of heat, frost and/or thaw cycles, and floods from water runoff.
- A series of adaptation and mitigation measures amounting to roughly $8 million was proposed to limit the risk of early failure.

Therefore, analysis with this new digital due diligence approach allows investments to be channeled to the most critical items and optimizes the overall financial performance of the project. From the initial data available on traffic estimations, geotechnical reports, material sources, and the base design, five alternative designs have been produced to find the best one in terms of the CFP, costs, natural resources consumption, and maintenance costs or service life. Digital due diligence focuses on each of those components and each of the design scenarios. The design can then be selected according to the integrated goals of the project and in terms of the CFP, natural resources consumption, budget, safety, and service life.

Carbon Dioxide Calculator for Afforestation Schemes

The carbon dioxide (CO_2) calculator is an Excel-based tool used in feasibility studies related to carbon offsetting of road projects in Central Asia for estimating the CO_2 absorbed by trees and shrubs, among others. The tool can be downloaded from the Asian Development Bank website. Below is the content of the "README" file provided with the tool in the download page, which describes the tool's capabilities.

A. Introduction

CO_2–Calculator – (v4.22) is an Excel-based calculation tool to be used in the preliminary assessment of carbon absorption caused by afforestation in road settings.

CO_2–Calculator – (v4.22) is compatible with the Excel 2007 and earlier versions for Windows.

B. Disclaimer

The tool is only intended for the preliminary assessment of the required number (or area) of certain tree species that could more or less efficiently sequestrate CO_2 during the lifetime of a project. An example of an application using this tool is predicting—at the feasibility stage—the required number and types of tree species to meet certain goals of CO_2 sequestration over a certain number of years of service of the afforestation scheme of the road, comparing ecological solutions, etc.

C. How the Tool Works

The spreadsheet has been designed using two alternative methods for the calculation of CO_2 absorption by vegetation and forests:

(i) Intergovernmental Panel on Climate Change (IPCC) (2003): Good Practice Guidance for Land Use, Land-Use Change and Forestry; and
(ii) United States Environmental Protection Agency (EPA) (1998): Method for Calculating Carbon Sequestration by Trees in Urban and Suburban Settings, April 1998, US Department of Energy, April.

There are some major differences between the two methods and the interested readers are referred to the original publications available online (free share). The substantial difference between the EPA and the IPCC methods lies in the fact that, while the EPA considers a certain number of separate tree species into a database, the IPCC method groups family of plants per geographic areas (mountains, floodplains, desert, etc.). Hence, while the IPCC method does not require to decide the specific tree type a priori, the EPA method requires to determine first the tree species intended for afforestation. Other main differences that may lead to sometimes very different results are based on assumptions for differential absorption over the life of the trees, including survival rate (EPA) and rate of growth (EPA).

For the interpretation of the output, when the EPA and the IPCC methods are considered simultaneously, some degree of judgment is required. For instance, the IPCC result, which is usually more conservative, could be the lower limit and the EPA result could be the upper limit of potential sequestration over the lifetime of the project. Sometimes, averaging the outputs may provide sufficient preliminary estimates.

D.   How to Use the Tool

When opening the file for the first time, the tool will typically look like the image below, which gives the option of calculations according to the EPA method (left, green) and the IPCC method (right, blue).

STEP 1. Define the project data.

Open the "project data" sheet and populate the rows with the information related to the project.

Project data	
Project name	Highway A380
Section	Kogon - Koson
Chainage	form km 200.3 to km 328.6
Username	Mr. Happy
Notes	Preliminary Assesment

Project-Data | CO2-Calculator | Report | Help

STEP 2. Calculate CO_2 absorption using the two methods.

EPA Method

Return to the CO_2-Calculator sheet. Choose whether the known variable is either the number of trees, or the area of the trees (in hectares) that will be planted (when the area is chosen first, the tool assigns by default a tree density to convert the area into a number of trees and vice versa). If the known variable is the number of trees, select "NUMBER OF TREES" and insert the number in the orange box. The example below considers 1,000 trees. Select the number of years of service of the afforestation scheme. The example below considers 10 years, meaning it considers the CO_2 absorption by the selected tree species over a period of 10 years. Note that the maximum years of service of the afforestation scheme for a project is 60 years, according to the parameters reported in Table 2 of the EPA method.

EPA Method		
Number of years considered	10	About
Input data — ○ Area (ha) ◉ Number of trees	1000	Set Report

Next, choose one tree species from the scroll down menu and press the "ADD DATA" button.

All tree species have also an expressed affinity in percentage (%) to the environment of Uzbekistan and bordering countries of Central Asia:

(i)   100% corresponds to species that are fully endemic to Central Asia,
(ii) 50% corresponds to Central Asia species not mentioned by the EPA but to which some plants in the EPA database may be compared to, and
(iii) 0% corresponds to "no affinity" to Central Asia.

In the example below, cedar-red tree is selected, which is 100% affine to Uzbekistan. The total CO_2 sequestration induced by this tree planted in 1,000 units is 38.60 tons over 10 years.

EPA Method. Number of years considered = 10   22/02/2022 - 13:03

Species Name	Number of trees planted	Area estimated (ha)	CO2 sequestred (t)
Cedar-red, eastern, Juniperus virginiana	1,000	1.60	38.60
Total	1,000	1.60	38.60

Next, another tree species is added to the cedar-red. One hundred hectares of walnut trees is selected for 10 years of service life. Repeating the previous steps, the output now consists of the sequestration by cedar-red and by walnut, both 100% affine to Central Asia. The result is reported in the example below.

EPA Method. Number of years considered = 10   22/02/2022 - 13:08

Species Name	Number of trees planted	Area estimated (ha)	CO2 sequestred (t)
Cedar-red, eastern, Juniperus virginiana	1,000	1.60	38.60
Walnut, black, Juglans nigra	62,500	100.00	6,548.08
Total	63,500	101.60	6,586.68

To remove one row, press the "DELETE LAST ROW" button. To delete all data, press the "DELETE ALL DATA" button.

Note: It is not possible to change the target service life while the contribution of trees is being considered from the scroll down menu. To change the target service life, rows must be emptied first, and a new calculation must be made after changing the number of years considered.

When the process is complete, press the "SET REPORT" button and a report will be automatically generated. Go to "PRINT" in the main Excel menu and select the "PRINT TO PDF" option to generate a PDF file of the calculation report as shown in the example below.

Calculation of CO_2 sequestration by trees

Project data	
Project name	Highway A380
Section	Kogon - Koson
Chainage	form km 200.3 to km 328.6
Username	Mr. Happy
Notes	Preliminary Assesment

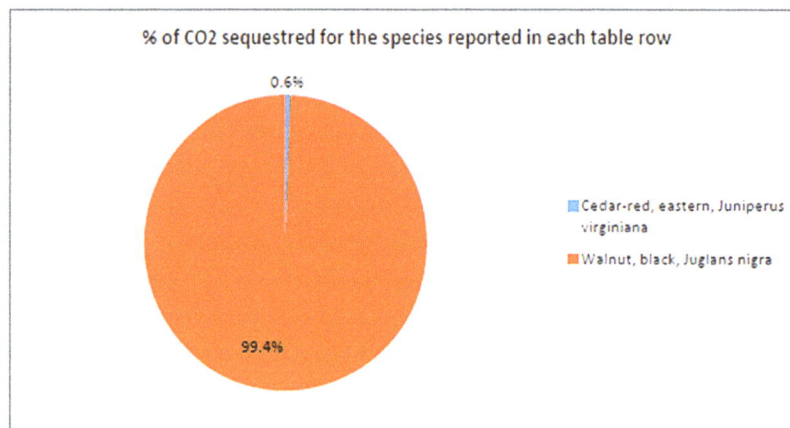

% of CO2 sequestred for the species reported in each table row

0.6%

99.4%

- Cedar-red, eastern, Juniperus virginiana
- Walnut, black, Juglans nigra

EPA Method. Number of years considered = 10			22/02/2022 - 13:12
Species Name	Number of trees planted	Area estimated (ha)	CO2 sequestred (t)
Cedar-red, eastern, Juniperus virginiana	1,000	1.60	38.60
Walnut, black, Juglans nigra	62,500	100.00	6,548.08
Total	63,500	101.60	6,586.68

IPCC Method

The tool works in the same manner as with the EPA version, except that the scroll-down menu does not populate with single tree species but with groups and related ecological environment. The example below shows the result of the calculations of CO_2 sequestration for the 20-year service life by 25,000 mountain forest trees (number is the input here, area is calculated) and by 100 hectares of shrubs (area is the input here, number is calculated).

IPCC Method

Number of years considered   20   About

Input data
◉ Area (ha)   ◯ Number of trees   100   Set Report

Species Name - (Natural zones)

Juniper arboral - (Mountain forests)
Other arboreal species growing in mountains - (Mountain forests)
Poplar (Asiatic poplar) - (Valley and floodplain forests)
Other arboreal species growing mainly in valleys and floodplains - (Valley and floodplain forests)
Saxaul - (Desert forests)
Shrubs

Add data

Delete last row

Delete all data

IPCC Method.   Number of years considered = 20   **22/02/2022 - 13:17**

Species Name	Number of trees estimated	Area (ha)	CO_2 sequestred (t)
Juniper arboral	25,000	40.00	91.29
Shrubs	62,500	100.00	3,220.53
Total	**87,500**	**140.00**	**3,311.81**

To generate a report, follow the steps reported for the EPA method.

Note: A report can also be generated with both the EPA and the IPCC methods by clicking on the REPORT sheet and by either creating a print screen or snip and snatch.

APPENDIX 2

Stiffness, Resilient Modulus, and Deformability of Soils

The assumption in pavement design is that materials behave like perfectly elastic materials—meaning, the relationship between applied stress σ and generated strain ε is of the following form:

$$\sigma = K\varepsilon$$

where K is the stiffness of the material, or the slope of the stress–strain linear relationship for a perfectly elastic material loaded in compression.[1] The slope, in a perfectly elastic material, does not vary irrespective of the stress applied. In nonelastic materials, on the other end, when stress is varied, the slope of the stress-strain curve varies. Figure A2.1 explains Hooke's law.

The equation can also be expressed as

$$\varepsilon = \frac{\sigma}{K}$$

Hence, soils with low stiffness have more deformability ε than stiffer soils, for the same applied stress σ. In pavement technology, more vertical strain in the soil equates to more damage to the pavement structure. Figure A2.2 shows the effect of excessive vertical deformation in the soil underlying an asphalt layer. Because the vertical strain on top of the subgrade is high as a consequence of reduced stiffness, the subgrade compresses and the upper layers (granular base and asphalt) will deflect as well, creating rutting and damage in the asphalt layers.

Materials like clay, silt, sand, and gravel used in road and railway construction to build embankments and pavement structure are generally unbound (or loose). This means that they have no tensile or compressive strength, and the only way they can resist deformation is by internal friction when the unbound particles slip past each other when the mass of the soil is loaded. When soil is loaded, its density increases by reducing the volume of air voids and by bringing the grains closer together. Compaction of soils ensures that they have reached the maximum packing of grains—the maximum density and the minimum volume of air voids. The unit weight of loose sandy gravel dumped on the ground from a truck may be about 1,700–1,800 kilograms (kg) per cubic meter (m^3). If a layer of this material is compacted to its maximum density, it may reach values in excess of 2,200 kg per m^3 as a consequence of air being replaced by more solid grains during the process.

When an element of soil is loaded under the wheel of a truck, for instance, the element will receive a vertical stress. It will want to compress but will be confined laterally by the surrounding soil (i.e., the surrounding soil will restrain the element in all directions). This causes a biaxial state of stress, as if the element is loaded both axially and laterally (see vertical and horizontal black arrows in Figure A2.3).

---

[1]   For instance, a cylinder-shaped or prism-shaped specimen being axially loaded.

Figure A2.1: Linear Elasticity

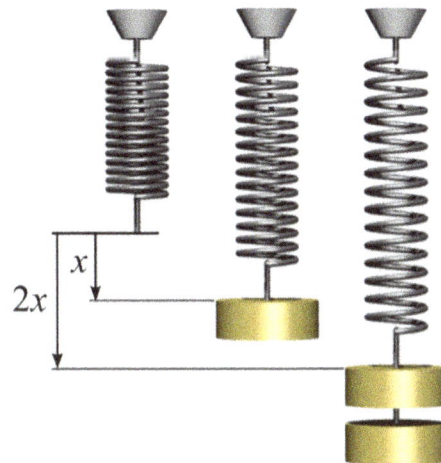

Note: By doubling the load (represented by the yellow brass rings) at one end of a spring while the opposite end is blocked, the initial displacement x (i.e., the elongation) is doubled to $2x$, meaning that there is a linear relationship between applied load and generated deformation. The ratio between the change in length caused by the applied load and the initial length of the spring at rest is called strain. The stiffness of the spring is equivalent to the stiffness K. When load is removed, the spring returns to its original length at rest. The only condition for this to happen anytime when the load is removed is that it does not exceed the yielding strength of the spring, which would cause it to elongate beyond the elastic point.

Source: Wikimedia – Hooke's Law Springs.

Figure A2.2: Rutting of Flexible Pavement

Note: Rutting is caused by traffic in the subgrade first and then it is reflected also in the upper asphalt concrete.
Source: Authors.

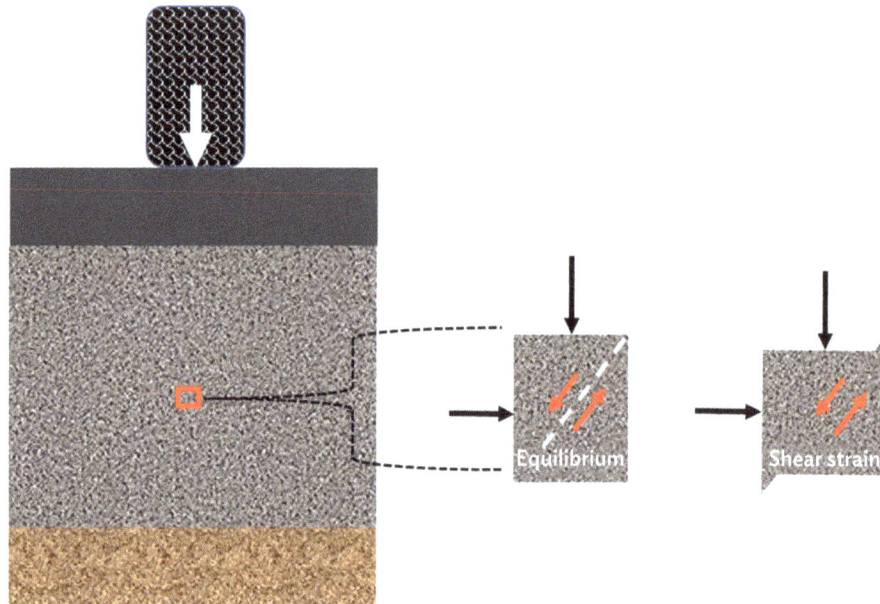

Figure A2.3: Shear Stress on a Soil Mass

Equilibrium

Shear strain

Note: Soils can only deform by shear, and in doing so, the slide along the planes inclines at an angle from the direction of the load.

Source: Authors.

The lateral stress is always smaller than the vertical stress,[2] and this causes shear stresses to form on a plane inclined to the direction of the vertical stress. If the vertical stress exceeds the resistance to sliding along this plane,[3] the element will fail in shear and it will deform by sliding (Figure A2.3). We call the difference between vertical stress σ_1 and horizontal stress σ_2 the deviatoric stress $\Delta\sigma = \left(\frac{\sigma_1 - \sigma_2}{2}\right)$. In Figure A2.3, this is represented by the red arrows on the inclined plane (dotted white line).

These conditions can be repeated in the laboratory by doing triaxial tests on soils. These measure the vertical deformation (or the axial strain ε) caused by the application of a vertical load from zero to the value σ_1 at a constant lateral pressure σ_2.

---

[2]   The ratio between vertical stress and lateral stress is the coefficient of earth pressure at rest (which may be assumed to be 0.5, meaning that for practical purposes, horizontal stress is roughly 50% of the applied vertical stress).

[3]   That is, if the shear stresses overcome the shear strength of the material.

This residual strain is called plastic or nonrecoverable strain, and the difference between the maximum strain measured at the peak stress and the plastic strain is the elastic or recoverable strain. If this cycle is repeated many times, it would become apparent that the amount of elastic strain increases with the number of load repetitions until it no longer increases, while the amount of plastic strain decreases until It becomes negligible. The ratio between the deviatoric stress and the axial elastic strain taken after many repetitions is the resilient modulus M_r, which is a measure of the stiffness of soils and granular materials in service.

$$M_r = \frac{deviatoric\ stress}{elastic\ strain\ after\ nth\ load\ repetitions}$$

From Stability to Sliding of Soil Masses Resting on Slopes

Debris flows and mudflows are the expression of a range of combinations of factors related to the nature of the ground, amount of water, and ground relief, among others. Only at the two extremes of the spectrum can they be classified as genuine debris flows or mudflows. Therefore, it is preferable to analyze the two phenomena as one. The triggering event for this gravity-induced phenomenon is the rapid saturation of the colluvium or of the bedrock after intense rain events. Intense and/or extreme rain events insisting on a specific area reduce the internal friction of materials near the surface through saturation and/or liquefaction and cause delamination, mobilization, or flow. Coarse, granular materials generally require steep slopes to mobilize and begin moving, while with dominant clay and silt, gravity flows can occur on slopes of a few degrees only.

An unloaded mass of soil of height H resting on a flat surface is subjected at its base to a vertical stress σ (Figure A3.1) given by

$$\sigma = \gamma H$$

where

γ = unit weight of soil at natural moisture content (kilonewton [kN]/cubic meter [m³])

H = height of soil (meter)

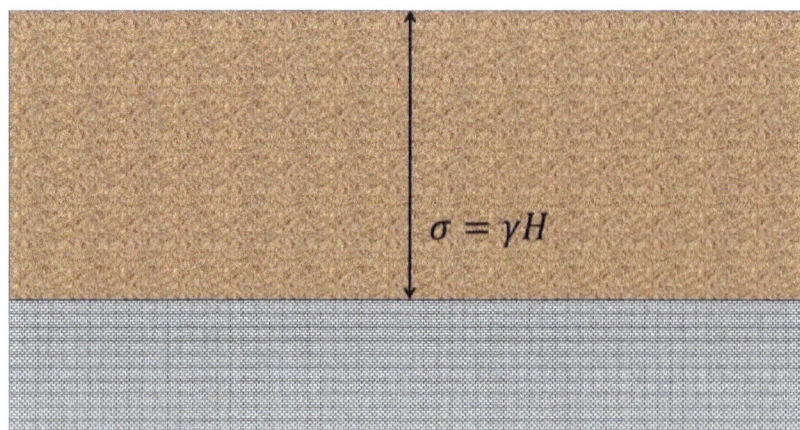

Figure A3.1: Vertical Stress on a Soil Mass

$$\sigma = \gamma H$$

Note: The vertical stress acting perpendicular to the base of the mass of soil height H is given by the product of the unit weight times the height.

Source: Authors.

Hence, on a flat surface, the only component of stress due to gravity is vertical. If the surface is inclined by an angle α, a component of stress parallel to the slope will arise, which will pull the mass of soil along the slope under the action of gravity. The value of the angle of the slope α for which the mass of soil begins sliding down is called the friction angle φ of the soil (Figure A3.2). Thus, the shear stress along the plane inclined at angle φ for which sliding begins is given, using elementary trigonometry, by

$$\tau = \gamma H tan\varphi = tan\varphi$$

Figure A3.2: Vertical and Shear Stresses on an Inclined Soil Mass

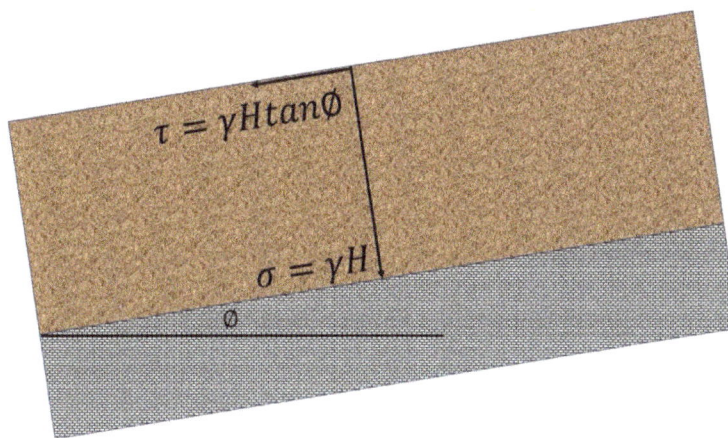

Note: When the mass of soil is inclined, gravity will act to push the mass toward the slope and to pull it down the slope.

Source: Authors.

Along with the component of shear stress parallel to the slope, the mass of soil is also subjected to a vertical stress perpendicular to it, which is caused by the contact pressure of each grain with the surrounding ones under the action of gravity. Whenever the soil is saturated with water, the buoyancy effect of water reduces the contact pressure at any depth by the value of the pore pressure at the same depth. This is the concept of effective vertical stress used in geotechnical engineering, which is equal to the difference between the vertical stress at depth minus the pore pressure at the same depth.

Finally, when a mass of very fine-grained soil, such as clay or silt, is placed on an inclined plane, it is observed that before the mass of soil slides down as the angle is increased, it is necessary to win another force—internal cohesion (c). Cohesion arises from capillary and electrostatic forces that work to bond very fine grains together. Cohesion is lost when very fine materials become saturated. Granular materials, such as sand and gravel, have no cohesion by definition regardless of the moisture content.

If a mass of soil begins sliding, it means that the equilibrium between resisting forces (pushing the mass *against* the slope) and acting forces (the force pulling the soil mass *along* the slope) is lost and there is a *failure of stability*.

The failure criterion leading to the mobilization and flow of materials (i.e., granular and cohesive; the only difference is that in granular soils, cohesion is zero) is governed by the following equation that condenses in symbolic and mathematical form introduced above:

$$\tau res = (\sigma - u)\tan\varphi + c$$

where

τres = resisting shear stress acting parallel to the slope because of gravity

σ = vertical stress acting perpendicular to the slope because of gravity

u = water pore pressure

$(\sigma - u)$ = effective vertical stress (as defined in geotechnical engineering)

c = cohesion (as defined in geotechnical engineering)

φ = angle of friction of the soil (as defined in geotechnical engineering)

The equation above says that the resisting shear stress to sliding is a function of the effective vertical stress and cohesion.

When water saturates the soil (i.e., during intense rain events), pore pressure increases and the effective vertical stress decreases. Also, cohesion, if originally present, is lost by saturation that breaks down the grain-to-grain capillary and electrostatic forces. However, the saturated material is also heavier than the dry material because the saturated material contains water instead of air to fill the pores, increasing the unit weight from γdry (dry unit weight) to γsat, which is the saturated unit weight.

This means that the saturated material will receive a larger shear pull along the slope because of gravity than a dry one, since a saturated soil always weighs more than a dry one. Hence, for a saturated mass of soil, the pulling shear stress $\tau pull$ will increase as a result of an increase in weight of the soil mass, while the resisting shear stress τres is reduced by water saturation.

Hence,

at equilibrium (rest): $\tau_{pull} = \gamma_{sat} H\tan\alpha \leq (\gamma_{sat} H\tan\varphi - u)\tan\varphi + c = \tau_{res}$

failure (sliding): $\tau_{pull} = \gamma_{sat} H\tan\alpha > (\gamma sat H\tan\varphi - u)\tan\varphi + c = \tau_{res}$

where α is the angle of the slope and φ is the soil's friction angle. Some values of the friction angles of common granular soils are given in Table A3.

Table A3: Typical Values of the Angle of Soil Friction

Type of Soil	Angle of Friction
Sand, round, loose	28°–30°
Sand, round, medium	30°–35°
Sand, round, dense	35°–38°
Sand, angular, loose	30°–35°
Sand, angular, medium	35°–40°
Sand, angular, dense	40°–45°
Sandy gravel	34°–48°

Source: Authors.

The equations above show that, if the unit weight is known, together with the water pressure, friction angle, and cohesion of the soil, the critical angle of the slope for sliding is simply given by

$$\alpha = arctan\left[\frac{(\gamma_{sat}Htan\varphi - u)tan\varphi + c}{\gamma_{sat}H}\right]$$

where the unknown parameter is the angle of the slope α to cause sliding.

Shear Strength of Rock Masses

The prevailing system of cracks or the sequence of strata of rocks can be oriented favorably or unfavorably to sliding and toppling, among others. Also, cracks, joints, and strata may be smooth, rough, filled with clay, or clean. The mechanics governing the movement of rock masses depends on the balance between a pulling force downhill and a resisting force due to friction mobilized within the rock mass.[1]

The equation below indicates that when joints are rough and no water or clay filling is present, the shear strength of the rock mass is larger than when joints are smooth and/or filled with water and clay. Also, for the shear pull to act, it is necessary that the planes of weakness of the rock mass have an orientation favorable to sliding in the first place.

$$\tau = \gamma H \tan \left[(JRC \, \log_{10} (JCS/(\gamma H) + \varnothing_b) \right]$$

where

γH = vertical stress on the plane of sliding

JRC = joint roughness coefficient (a quantity that depends on how smooth or rough a joint is)

JCS = joint compressive strength (equal to the uniaxial compressive strength of the intact rock, or to 75% of this value if the rock is weathered)

$\varnothing_b$ = basic friction angle (0° for perfectly smooth joint, 20° for rough joints)

---

[1]   F. G. Bell. 2007. *Engineering Geology*. Elsevier.

APPENDIX 5
Stabilizers for the Engineering Modification of Soils

To turn a low stiffness material into a high stiffness one, several types of binding agents may be used, and the choice depends on the nature of the soil and the desired mechanical properties in service. Below is a brief review of common binders and applications (Table A5 presents an explanation of the terms used).

Portland cement can be used effectively as a stabilizer for a wide range of materials. However, the soil should generally have a plasticity index of less than 30 or the stabilization effects will be reduced. For coarse-grained soils, the amount passing the 4.75-millimeter (mm) sieve should be preferably greater than 45% or the stabilization effects will be reduced. The amount of cement used depends on the stiffness required, but it normally comprises between 90 and 180 kilograms per cubic meter of stabilized material. Stiffness is usually evaluated indirectly through compressive strength. Generally, compressive strength of good quality stabilized soils with cement should not exceed 10 megapascals to avoid producing a layer that is too brittle and, thereby, prone to cracking.

Experience shows that **lime** will react with many medium-, moderately fine-, and fine-grained soils to produce decreased plasticity, increased workability, reduced swelling, and increased strength. Lime can successfully stabilize soils classified according to the Unified Soil Classification System (USCS) as CH, CL, MH, ML, OH, OL, SC, SM, GC, GM, SW-SC, SPSC, SM-SC, GW-GC, GP-GC, ML-CL, and GM-GC. Lime is considered with all soils having a plasticity index greater than 12 and more than 25% of the soil passing the 0.075 mm sieve. Stabilization with lime provides generally low compressive strength and high ductility (low cracking potential). Lime has the same carbon footprint with that of Portland cement.

Fly ash, when mixed with lime or cement, can effectively stabilize most coarse- and medium-grained soils; however, the plasticity index cannot be greater than 25 and remain effective. Soils classified by the USCS as SW, SP, SP-SC, SW-SC, SW-SM, GW, GP, GP-GC, GW-GC, GP-GM, GW-GM, GC-GM, and SC-SM can be stabilized with fly ash. Stabilization with large volume of fly ash provides low compressive strength and high ductility (low cracking potential). Fly ash has a negligible carbon footprint (about 50 times lower) in comparison with that of Portland cement.[1]

Most **bitumen** soil stabilization is performed with asphalt cement and asphalt emulsions. Soils that can be stabilized effectively with bituminous materials usually contain less than 30% passing the 0.075 mm sieve, and have a plasticity index of less than 10%. Soils classified by the USCS as SW, SP, SW-SM, SP-SM, SW-SC, SP-SC, SM, SC, SMSC, GW, GP, SW-GM, SP-GM, SW-GC, GP-GC, GM, GC, and GM-GC can be effectively stabilized with bituminous materials, provided the above mentioned gradation and plasticity requirements are met. Stabilization with bitumen provides low compressive strength and high ductility (low cracking potential).

---

[1] This is because fly ash is a by-product of another industrial process, meaning that the carbon released to produce electric energy from a coal plant is not going to be increased further when fly ash, which forms when coal is burned, is recycled as a cementitious material.

Combinations of lime and cement and/or fly ash are finally often acceptable to obtain combined properties of the material. Lime can be added to the soil to increase the soil's workability and mixing characteristics as well as reduce its plasticity. Cement or fly ash can then be mixed into the soil to provide rapid strength gain. Combinations of lime and asphalt are often acceptable stabilizers. The classifications of soils for engineering purposes are shown in Table A5.

Table A5: Classification of Soils for Engineering Purposes Using the Unified Soil Classification System

Major Divisions		Group Symbol	Group Name
Coarse-grained soils (>50% retained on the 0.075 mm sieve)	Clean gravel (<5% passing the 0.075 mm sieve)	GW	Well-graded gravel
		GP	Poorly graded gravel
	Gravel with fines	GM	Silty gravel
		GC	Clayey gravel
	Clean sand	SW	Well-graded sand
		SP	Poorly graded sand
	Sand with fines	SM	Silty sand
		SC	Clayey sand
Fine-grained soils (>50% passing the 0.075 mm sieve)	Inorganic	ML	Silt
		CL	Clay

mm = millimeter.
Source: Authors.

www.ingramcontent.com/pod-product-compliance
Lightning Source LLC
Chambersburg PA
CBHW061221270326
41926CB00032B/4799